T0178600

Lecture Notes in Computer Science　13100

Founding Editors

Gerhard Goos
Karlsruhe Institute of Technology, Karlsruhe, Germany

Juris Hartmanis
Cornell University, Ithaca, NY, USA

Editorial Board Members

Elisa Bertino
Purdue University, West Lafayette, IN, USA

Wen Gao
Peking University, Beijing, China

Bernhard Steffen ⓘ
TU Dortmund University, Dortmund, Germany

Gerhard Woeginger ⓘ
RWTH Aachen, Aachen, Germany

Moti Yung ⓘ
Columbia University, New York, NY, USA

More information about this subseries at https://link.springer.com/bookseries/7409

Mantas Šimkus · Ivan Varzinczak (Eds.)

Reasoning Web

Declarative Artificial Intelligence

17th International Summer School 2021
Leuven, Belgium, September 8–15, 2021
Tutorial Lectures

 Springer

Editors
Mantas Šimkus 🆔
TU Wien
Vienna, Austria

Ivan Varzinczak 🆔
Université d'Artois and CNRS
Lens, France

ISSN 0302-9743   ISSN 1611-3349   (electronic)
Lecture Notes in Computer Science
ISBN 978-3-030-95480-2   ISBN 978-3-030-95481-9   (eBook)
https://doi.org/10.1007/978-3-030-95481-9

LNCS Sublibrary: SL3 – Information Systems and Applications, incl. Internet/Web, and HCI

© Springer Nature Switzerland AG 2022
This work is subject to copyright. All rights are reserved by the Publisher, whether the whole or part of the material is concerned, specifically the rights of translation, reprinting, reuse of illustrations, recitation, broadcasting, reproduction on microfilms or in any other physical way, and transmission or information storage and retrieval, electronic adaptation, computer software, or by similar or dissimilar methodology now known or hereafter developed.
The use of general descriptive names, registered names, trademarks, service marks, etc. in this publication does not imply, even in the absence of a specific statement, that such names are exempt from the relevant protective laws and regulations and therefore free for general use.
The publisher, the authors and the editors are safe to assume that the advice and information in this book are believed to be true and accurate at the date of publication. Neither the publisher nor the authors or the editors give a warranty, expressed or implied, with respect to the material contained herein or for any errors or omissions that may have been made. The publisher remains neutral with regard to jurisdictional claims in published maps and institutional affiliations.

This Springer imprint is published by the registered company Springer Nature Switzerland AG
The registered company address is: Gewerbestrasse 11, 6330 Cham, Switzerland

Preface

The Reasoning Web (RW) series of annual summer schools has become the prime educational event in reasoning techniques on the Web. Since its initiation in 2005 by the European Network of Excellence (REWERSE), RW has attracted both young and established researchers. As with the previous edition of RW, this year's school was part of Declarative AI (https://declarativeai2021.net), which brought together the 5th International Joint Conference on Rules and Reasoning (RuleML+RR 2021), DecisionCAMP 2021, and the 17th Reasoning Web Summer School (RW 2021). As a result of the COVID-19 pandemic, Declarative AI 2021 was held as an online event.

This year's school covered various aspects of ontological reasoning and related issues of particular interest to Semantic Web and Linked Data applications. The invitations to teach at the summer school as well as to submit lectures for publication were carefully vetted by the Scientific Advisory Board, consisting of six renowned experts of the area. The following eight lectures were presented during the school (further details can be found at https://declarativeai2021.net/reasoning-web):

1. Foundations of graph path query languages
   by Diego Figueira (CNRS, France)
2. On combining ontologies and rules
   by Matthias Knorr (Universidade Nova de Lisboa, Portugal)
3. Modelling symbolic knowledge using neural representations
   by Steven Schockaert and Victor G. Basulto (Cardiff University, UK)
4. Mining the Semantic Web with machine learning: main issues that need to be known
   by Claudia d'Amato (University of Bari, Italy)
5. Belief revision and ontology repair
   by Renata Wassermann (University of São Paulo, Brazil)
6. Temporal ASP: from logical foundations to practical use with telingo
   by Pedro Cabalar (University of A Coruña, Spain)
7. SHACL: from data validation to schema reasoning for RDF graphs
   by Paolo Pareti (University of Winchester, UK)
8. Explanations in data management and classification in machine learning via counterfactual interventions specified by answer-set programs
   by Leopoldo Bertossi (Adolfo Ibáñez University, Chile)

The present volume contains lecture notes complementing most of the above lectures. They are meant as accompanying material for the students of the summer school in order to deepen their understanding and serve as a reference for further detailed study. All articles are of high quality and have been peer-reviewed by members of the Scientific Advisory Board as well as additional reviewers.

We want to thank everybody who helped make this event possible. Since teaching is the main focus of a summer school, we first thank all the lecturers; their hard work and commitment ensured a successful event. We are also thankful to all members of the Scientific Advisory Board; their timely feedback concerning the technical program and submitted lecture notes helped us organize a high-quality event. Finally, we want to express our gratitude to the organizers of Declarative AI 2021 and those of the previous edition of RW for their constant support. The work of Mantas Šimkus was supported by the Vienna Business Agency and the Austrian Science Fund (FWF) projects P30360 and P30873.

October 2021

Mantas Šimkus
Ivan Varzinczak

Organization

Program Chairs

Mantas Šimkus   TU Wien, Austria
Ivan Varzinczak   Université d'Artois and CNRS, France

General Chair

Jan Vanthienen   KU Leuven, Belgium

Scientific Advisory Board

Thomas Eiter   TU Wien, Austria
Birte Glimm   University of Ulm, Germany
Domenico Lembo   Sapienza University of Rome, Italy
Marco Manna   University of Calabria, Italy
Andreas Pieris   University of Edinburgh, UK
Anni-Yasmin Turhan   TU Dresden, Germany

Additional Reviewers

Shqiponja Ahmetaj   TU Wien, Austria
Marco Calautti   University of Trento, Italy
Daria Stepanova   Bosch Center for Artificial Intelligence, Germany

Contents

Foundations of Graph Path Query Languages
Course Notes for the Reasoning Web Summer School 2021

Diego Figueira[(✉)]

Univ. Bordeaux, CNRS, Bordeaux INP, LaBRI, UMR 5800,
33400 Talence, France
`diego.figueira@labri.fr`

Abstract. We survey some foundational results on querying graph-structured data. We focus on general-purpose navigational query languages, such as regular path queries and its extensions with conjunctions, inverses, and path comparisons. We study complexity, expressive power, and static analysis. The course material should be useful to anyone with an interest in query languages for graph structured data, and more broadly in foundational aspects of database theory.

A graph database is an umbrella term for describing semi-structured data organized by means of *entities* (*i.e.*, nodes) and *relations* (*i.e.*, edges) between these entities. In other words, as a *finite graph*, which emphasizes the holistic, topological aspect of the model, where there is no order between nodes or edges. This is a flexible format, usually with no 'schemas', where adding or deleting data (or even integrating different data sources) does not imply rethinking the modeling of data. Data can be typically stored both in nodes and edges, but the shape of the graph itself is an essential part of the data. Querying mechanisms on this kind of data focus on the *topology* of the underlying graph as well as in the data contained inside the edges and nodes. This flexibility comes at a cost, since relations between entities have to be found in a possibly complex topology, most notably as *paths* or sets of paths in some specific configuration. Indeed a *path* in a graph database can be then seen as a first-order citizen. The most basic querying mechanism is then the problem of finding a "pattern" in the database, given as nodes and paths relating them with certain properties. This is, precisely, the kind of languages we will survey here, sometimes called "path query languages".

Example 1. Consider, for example, a very basic database of academic staff. This can be seen as a graph database, as shown in Fig. 1. The kind of queries we're interested in are those which exploit the topology of graph, such as "are there two persons with the same supervisor at friend distance at most 5?" or "find all pairs of co-authors with a common ancestor in the supervisor-relation".   ◁

© Springer Nature Switzerland AG 2022
M. Šimkus and I. Varzinczak (Eds.): Reasoning Web 2021, LNCS 13100, pp. 1–21, 2022.
https://doi.org/10.1007/978-3-030-95481-9_1

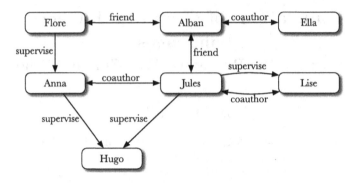

Fig. 1. A simple bibliometric graph database.

Graph databases are relevant to a growing number of applications in areas such as the Semantic Web, knowledge representation, analysis of Social Networks, biological and scientific databases, and others. This data model encompasses formats such as RDF [38], or property graphs. This is why in the last years there has been many theoretical and practical developments for querying graph databases (see [1,2,6,45] for surveys).

One of the most important research trends has hinged on the development of graph query languages that can reason about *topological* aspects of the graph. They are also known as *path query languages*, because topological information in the database typically amounts to querying the existence of paths satisfying certain constraints. The most basic form of navigation consists of querying whether there is a path with a certain property between two nodes. This type of queries have been introduced as Regular Path Queries, or RPQ [40], and it has laid the foundations of many more expressive query languages, including Conjunctive Regular Path Queries (CRPQ) [28] or Extended CRPQ (ECRPQ) [9].

Outline. This brief survey concerns the computational task of querying graph databases via navigational query languages. We focus on the language of *regular path queries* and its standard extensions. We study the complexity of evaluation and static analysis tasks, and its expressive power.

1   Preliminaries

We will assume familiarity with some basic automata theory notions such as non-deterministic finite automata (NFA), regular languages, regular expressions and its paradigmatic problems of containment, emptiness and equivalence. We use $\mathbb{A}, \mathbb{B}$ to denote finite alphabets. In our examples, we use the standard syntax for regular expressions over a finite alphabet $\mathbb{A}$

$$regexp ::= \emptyset \mid \epsilon \mid a \mid regexp \cdot regexp \mid regexp + regexp \mid regexp^* \mid regexp^+ \qquad a \in \mathbb{A}$$

with the semantics $\llbracket\ \rrbracket : regexp \to 2^{\mathbb{A}^*}$

$$\llbracket \emptyset \rrbracket = \emptyset, \quad \llbracket \epsilon \rrbracket = \{\epsilon\}, \quad \llbracket a \rrbracket = \{a\}, \quad \llbracket e_1 + e_2 \rrbracket = \llbracket e_1 \rrbracket \cup \llbracket e_2 \rrbracket,$$

$$\llbracket e^+ \rrbracket = \{u_1 \cdots u_n : n \geq 1 \text{ and } u_i \in \llbracket e \rrbracket \text{ for every } i\},$$

$$\llbracket e^* \rrbracket = \{\epsilon\} \cup \llbracket e^+ \rrbracket, \quad \llbracket e_1 \cdot e_2 \rrbracket = \{u \cdot v : u \in \llbracket e_1 \rrbracket, v \in \llbracket e_2 \rrbracket\}.$$

We use the word orderings of

- **prefix:** u is a prefix of v if $v = u \cdot w$ for some w;
- **suffix:** u is a suffix of v if $v = w \cdot u$ for some w;
- **factor** (*a.k.a.* infix, subword): u is a factor of v if $v = w \cdot u \cdot w'$ for some w, w';
- **subsequence** (*a.k.a.* scattered subword): u is a subsequence of v if u is the result of removing some (possibly none) positions from v.

We also use its "proper" versions: u is a **proper prefix** of v if it is a prefix of v and $u \neq v$; and similarly for the other orderings.

We also assume an elementary understanding of some fundamental complexity classes such as PTIME, NL, PSPACE, EXPSPACE, the polynomial hierarchy, etc.

We often blur the distinction between an NFA $\mathcal{A}$ over $\mathbb{A}$ and the language $L(\mathcal{A}) \subseteq \mathbb{A}^*$ it recognizes; and we do similarly for regular expressions. In the sequel we may hence write $w \in c^* \cdot (a + b)^*$ or $w \in \mathcal{A}$. We denote by ϵ the empty word. We also assume some familiarity with the query language of Conjunctive Queries (CQ) and Unions of CQ (UCQ).

Graph Databases. We consider a **graph database** over a finite alphabet $\mathbb{A}$ to be a finite edge-labelled directed graph $G = (V, E)$ over a finite set of labels $\mathbb{A}$, where V is a finite set of vertices and $E \subseteq V \times \mathbb{A} \times V$ is the set of labelled edges. We write $u \xrightarrow{a} v$ to denote an edge $(u, a, v) \in E$. It should be stressed that this is often an *abstraction* for formats such as RDF [38] or property graphs (adopted, *e.g.*, by Neo4j). For example, the patterns used in SPARQL [32] (the W3C query language for RDF) are *triplets* rather than *edges*, but this can often be abstracted away by means of extra vertices and edges, without much loss of generality. Also, for most graph database formats, $\mathbb{A}$ may be from a complex infinite domain, and further nodes may be labelled also with data. Graph databases, as defined here, are then a basic abstraction of these models which allows us to focus on querying the *topology of the graph*.

A (directed) **path** π of length $n \geq 0$ in G is a (possibly empty) sequence of edges of G of the form $(v_0, a_1, v_1), (v_1, a_2, v_2), \ldots, (v_{n-1}, a_n, v_n)$. There is always an empty path starting and ending at the same node. The **label** $label(\pi)$ of π is the word $a_1 \cdots a_n \in \mathbb{A}^*$. When $n = 0$ the label of π is the empty word ϵ.

2   Conjunctive Regular Path Queries

In graph databases, a fundamental querying mechanism is based on the existence of some paths in the database with certain properties. These properties include that the label of a path must belong to a certain language, or that the starting or terminal vertices of some paths must be equal. This gives rise to the much studied class of *Regular Path Queries* (RPQ) and *Conjunctive Regular Path Queries* (CRPQ) [21].

Example 2. An example of a CRPQ query is

$$Q_1(x) = x \xrightarrow{a^*b} y \wedge x \xrightarrow{(a+b)^*c} y.$$

It outputs all vertices v having one outgoing path with label in a^*b and one outgoing path with label in $(a+b)^*c$. Further these paths must end at the same vertex.   ◁

Conjunctive Regular Path Queries (CRPQ) can be understood as the generalization of conjunctive queries with a very simple form of recursion. CRPQ are part of SPARQL, the W3C standard for querying RDF data [38], including well known knowledge bases such as DBpedia and Wikidata. In particular, RPQs are quite popular for querying Wikidata. They are used in over 24% of the queries (and over 38% of the unique queries), according to recent studies [16,37]. More generally, CRPQ constitute a basic building block for query languages on graph-structured data [6].

A **Regular Path Query (RPQ)** over the alphabet $\mathbb{A}$ is a query of the form

$$Q(x, y) = x \xrightarrow{L} y \tag{1}$$

where L is a regular language over $\mathbb{A}$, specified either as an NFA or a regular expression (we will not make a distinction here). Given a graph database G and a pair of node (v, v') therein, we say that the pair (v, v') **satisfies** Q if there exists a path π from v to v' such that $label(\pi) \in L$. The result of evaluating Q on G is then the set of all pairs (v, v') of G satisfying Q.

Example 3. Consider the RPQ

$$Q(x, y) = x \xrightarrow{\text{coauthor}^*} y.$$

It retrieves all pairs persons related by a coauthorship. In particular on the graph database defined in Example 1 it retrieves (Anna, Lise), among other pairs.   ◁

A conjunctive regular path query (CRPQ) is the closure under projection (*i.e.*, existential quantification) and conjunction of RPQ queries. That is, CRPQ is to RPQ what Conjunctive Queries is to first-order atoms. Concretely, a **Conjunctive Regular Path Query (CRPQ)** is a query of the form

$$Q(x_1, \ldots, x_n) = A_1 \wedge \cdots \wedge A_m$$

where the atoms $A_1, \ldots, A_m$ are RPQ. We call the variables $x_1, \ldots, x_n$ occurring on the left-hand side the **free variables**. Each free variable x_j has to occur also in some atom on the right-hand side, but not every variable on the right-hand side needs to be free.

A **homomorphism** from a CRPQ Q as above to a graph database $G = (V, E)$ is a mapping μ from the variables of Q (free and non-free) to V. Such a homomorphism **satisfies** an RPQ $A(x, y)$ if $(\mu(x), \mu(y))$ satisfies A; and it satisfies Q if it satisfies every RPQ atom of Q. The set of **answers** $Q(G)$ of a CRPQ $Q(x_1, \dots, x_n)$ over a graph database G is the set of tuples $(v_1, \dots, v_n)$ of nodes of G such that there exists a satisfying homomorphism for Q on G that maps x_i to v_i for every $1 \leq i \leq n$. We say that a CRPQ is **Boolean** if it has no free variables, in which case $Q(G) = \{()\}$ (where () denotes the empty tuple) if there exists a satisfying homomorphism or $Q(G) = \{\}$ otherwise. We often write $G \models Q$ instead of $Q(G) = \{()\}$. Most of the results we will present hold also for expressive extensions of CRPQ with finite unions and two-way navigation, known as UC2RPQ [18]. However, for simplicity of presentation, we will focus on CRPQ.

Example 4. Consider the (Boolean) CRPQ

$$Q_1() = x \xrightarrow{\text{supervise}^+} x$$

It checks if the supervisor relation has cycles (*i.e.*, it is true whenever there are). Another CRPQ could be

$$Q_2(x) = x \xrightarrow{\text{supervise}^+} y \wedge x \xrightarrow{\text{friend}} y$$

retrieving all persons being friends with some descendant in the supervisor genealogy.   ◁

It is worth observing that in the context of graph databases, a **Conjunctive Query (CQ)** is a CRPQ whose every regular expression denotes a language of the form $\{a\}$ for some $a \in \mathbb{A}$. Thus, CQ is included in CRPQ in terms of expressive power.

Alternative Semantics. For some applications such as transportation problems or DNA matching (see [5] for a more complete list of application scenarios) there is a need to require that the considered paths have no repeated nodes or no repeated edges. In this way, alternative semantics arise if we change the definition of "satisfaction" of an RPQ atom $x \xrightarrow{L} y$ for a given homomorphism μ. In the default (*a.k.a.* **arbitrary path**) semantics, we ask for the existence of any (directed) path from $\mu(x)$ to $\mu(y)$ with $label(\pi) \in L$. In the **trail** semantics, we demand that the path has also no repeated edges, and in the **simple path** semantics, we further enforce that the path must be simple (*i.e.*, no repeating vertices). It then follows that if $\bar{x} \in Q(G)$ under simple path semantics, then $\bar{x} \in Q(G)$ under trail semantics; and if $\bar{x} \in Q(G)$ under trail semantics then $\bar{x} \in Q(G)$ under arbitrary path semantics. But the converse directions do not hold in general. In the sequel we assume that we work with the default (*i.e.*, arbitrary path) semantics unless otherwise stated.

Structural Fragments of CRPQ. A standard way to define fragments of conjunctive queries is via their underlying graph (*a.k.a.* Gaifman graph). In a similar way, one can define fragments of CRPQ via its *underlying multi-graph*. Concretely, for any set X, let $\wp_2(X)$ denote the set of non-empty subsets of X of size at most 2. The **underlying multi-graph** of a CRPQ Q is the directed multi-graph (V, E, ν) where: V is the set of variables of Q, E is the set of atoms of Q, and $\nu : E \to \wp_2(V)$ is defined as $\nu(x \xrightarrow{L} y) = \{x, y\}$ for every RPQ atom $x \xrightarrow{L} y$ in Q. For a given class $\mathcal{C}$ of multi-graphs, let $\mathrm{CRPQ}(\mathcal{C})$ be the set of CRPQ whose underlying multi-graph is in $\mathcal{C}$.[1] In the sequel we will rather use the term *graph* to denote the underlying multi-graph of a CRPQ.

Example 5. Consider, for example, the CRPQ

$$Q(x, z) = x \xrightarrow{a^*} y \wedge y \xrightarrow{a+b^*} y \wedge x \xrightarrow{b^*} z \wedge z \xrightarrow{(b+c)^*} x$$

In Fig. 2 there is its graphic representation and its underlying graph.

◁

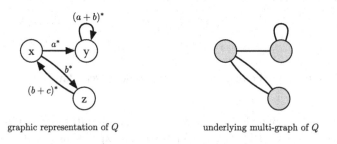

graphic representation of Q underlying multi-graph of Q

Fig. 2. Underlying graph of a CRPQ.

3   Evaluation of CRPQ

The evaluation problem is the most fundamental decision problem on databases: the problem of whether a given data is retrieved by a query on a database.

PROBLEM	Evaluation problem for a class $\mathcal{Q}$ of (graph database) queries (EVAL-$\mathcal{Q}$)
GIVEN	$Q \in \mathcal{Q}$, a graph database G, a tuple $\bar{x}$ of nodes
QUESTION	Is $\bar{x} \in Q(G)$?

---

[1] Why multi-graphs and not just graphs? It turns out that, contrary to what happens to Conjunctive Queries, the multiplicity of edges makes a difference for some problems, such as the containment problem. We need, hence, to have a more fine-grained notion than a simple graph.

Observe that the evaluation problem has two kind of inputs of very different nature: the query and the database. In terms of size, one should expect the query to be several orders of magnitude smaller than the database, which raises the question of, for example, whether different algorithms running in time $O(2^{|Q|} \cdot |D|)$, $O(|Q| \cdot 2^{|D|})$ or $O(|D|^{|Q|})$ should be justly placed in the same "complexity class". This is the reason why several complexity variants are often considered, useful to understanding the various aspects of the complexity for the evaluation problem. The default one is the **combined complexity**, where one considers both the query and database as being part of the input. The complexity when one considers the input query Q to be of *constant size* it is the **data complexity** [44]. Hence, an algorithm running in $O(2^{|Q|} \cdot |D|)$ would have exponential combined complexity but linear data complexity. If, on the other hand, one considers the database D to be of constant size we obtain the **query complexity**. There is, on the other hand, the **parameterized complexity** version of this problem in which the 'parameter' is the query, we will not give details here about parameterized complexity, and we refer the interested reader to [29]. On the parameterized complexity, the classes 'FPT' (for Fixed Parameter Tractable) and 'W[1]' are often considered as the PTIME and NP analog of classical non-parameterized complexity classes, respectively. The idea is that an algorithm is FPT if it runs in time $O(f(|Q|) \cdot |D|^c)$ for any computable f and constant c. Thus, an algorithm running in time $O(|D|^{|Q|})$ is not (in principle) FPT, but an algorithm running in time $O(2^{|Q|} \cdot |D|)$ is FPT.

Theorem 1 (Folklore). EVAL-*CRPQ is*

- NP-*complete in combined complexity,*
- NL-*complete is data complexity,*
- NP-*complete in query complexity,*
- W[1]-*complete in parameterized complexity.*

EVAL-*RPQ is*

- NL-*complete in combined complexity,*
- NL-*complete in data complexity,*
- NL-*complete in query complexity,*
- FPT *in parameterized complexity.*

That is, the combined complexity follows the same behavior as that of Conjunctive Queries, with the exception that evaluating the 'atoms' is an NL-complete task —essentially, the classical graph problem of existence of a source to target path. In other words, the lower bounds for combined and parameterized complexities follow from the following classical result for CQ.

Theorem 2. [19] EVAL-*CQ is*

- NP-*complete in combined complexity,*
- *in* LOGSPACE *(and in* AC^0*) in data complexity,*
- NP-*complete in query complexity,*
- W[1]-*complete in parameterized complexity.*

As discussed before, a standard way to define subclasses of CRPQ is by means of its underlying graphs. In the light of the results of Theorem 1 above, one natural concern is whether the combined and parameterized complexities can be improved by considering queries of some 'simple' structure. The question is then: given a class $\mathcal{C}$ of graphs, is EVAL-CRPQ($\mathcal{C}$) tractable? Or rather: For which $\mathcal{C}$ is EVAL-CRPQ($\mathcal{C}$) tractable?

As it turns out, by straightforward reductions to and from the Conjunctive Query case, we obtain that the RPQ complexity extends to any class of CRPQ defined by a bounded treewidth class.[2] This notion, in fact, characterizes the tractable complexity classes.

Theorem 3 (consequence of [31]). *Assuming* $W[1] \neq FPT$, *for any class* $\mathcal{C}$ *of graphs the following are equivalent:*

- *EVAL-CRPQ($\mathcal{C}$) is in polynomial time in combined complexity,*
- *EVAL-CRPQ($\mathcal{C}$) is FPT in parameterized complexity,*
- *$\mathcal{C}$ has bounded treewidth.*

The tractable cases of evaluation has been also extended to larger classes, in which either queries need to be equivalent to queries of bounded treewidth (obtaining FPT tractability) or they have to be homomorphically equivalent[3] to queries of bounded treewidth (obtaining polynomial time tractability) [42].

Alternative Semantics. Under alternative semantics, things are more complex, since EVAL-RPQ is already an NP-complete problem.

Theorem 4. *EVAL-RPQ is NP-complete both under trail and simple path semantics. Both in data and in combined complexity.*

In fact, NP-completeness under simple path or trail semantics already holds if we fix the query to be $x \xrightarrow{(aa)^*} y$ or $x \xrightarrow{a^*ba^*} y$ [40]. Interestingly, both these semantics enjoy a trichotomy characterization in terms of data complexity: for any fixed query $Q = x \xrightarrow{L} y$ the evaluation problem for Q is either NP-complete, NL-complete, or in LOGSPACE (even in AC^0, the data complexity of evaluating first-order formulas). What is more, given a query Q, one can effectively decide in which of these three cases falls (for each semantics).

Theorem 5 ([5,39]). *For each fixed regular language* $L \subseteq A^*$ *and for each* $\star \in \{\text{simple-path}, \text{trail}\}$, *the data complexity of EVAL-RPQ for* $x \xrightarrow{L} x$ *under $\star$-semantics is either NP-complete, NL-complete or in AC^0. Further, these characterizations are effective (and different for each semantics).*

---

[2] Intuitively, a graph with small treewidth resembles a tree (*e.g.*, trees have treewidth 1 and cacti have treewidth 2). Many results for trees can be generalized to bounded treewidth classes.

[3] For some suitable notion of homomorphism between queries.

4   Containment for CRPQ

As databases become larger, reasoning about queries (*e.g.*, for optimization) becomes increasingly important. One of the most basic static analysis problems on monotone query languages is that of query containment: *is every result returned by query Q_1 also returned by query Q_2, for every database?* This can be a means for query optimization, as it may avoid evaluating parts of a query, or reduce and simplify the query with an equivalent one. It falls in what is commonly known as *query reasoning* or *static analysis*, since it involves reasoning only about the query, and it may give rise to optimization tasks that can be carried out at compile time (rather than at running time). Furthermore, query containment has proven useful in knowledge base verification, information integration, integrity checking, and cooperative answering [18].

Concretely, given two CRPQ Q_1, Q_2, we say that Q_1 is **contained** in Q_2, denoted by $Q_1 \subseteq Q_2$, if $Q_1(G) \subseteq Q_2(G)$ for every graph database G, which raises the following decision problem for any fragment $\mathcal{Q}$ of CRPQ.

PROBLEM Containment problem for a class $\mathcal{Q}$ of (graph database) queries (CONT-$\mathcal{Q}$)
GIVEN $Q_1, Q_2 \in \mathcal{Q}$
QUESTION Is $Q_1(G) \subseteq Q_2(G)$ for every graph database G?

We say Q_1 is **equivalent** to Q_2, denoted by $Q_1 \equiv Q_2$, if $Q_1 \subseteq Q_2$ and $Q_2 \subseteq Q_1$.

The containment problem for RPQ and CRPQ are decidable in PSPACE and EXPSPACE respectively.

Theorem 6 (Folklore). CONT-*RPQ is* PSPACE-*complete.*

In fact, for any two RPQ $Q_1 = x \xrightarrow{L_1} y$ and $Q_2 = x \xrightarrow{L_2} y$ it is easy to see that $Q_1 \subseteq Q_2$ if, and only if, $L_1 \subseteq L_2$. Hence RPQ containment is reducible from and to language containment. Since regular language containment is a PSPACE-complete problem, it follows that CONT-RPQ is PSPACE-complete. On the other hand, the bounds for CRPQ are somewhat more involved.

Theorem 7 ([18,28]). CONT-*CRPQ is* EXPSPACE-*complete.*

It is interesting to remark that the above hardness result holds even for containment of CRPQ $Q_1 \subseteq Q_2$ where Q_1 is of the form $Q_1 = x \xrightarrow{L} y$ and Q_2 is of the form $Q_2 = \bigwedge_i x \xrightarrow{L_1} y$. In other words, CONT-CRPQ($\mathcal{C}$) is already EXPSPACE-hard for the class $\mathcal{C}$ of multigraphs having exactly two nodes, and even for Boolean queries.

However, in certain circumstances, the EXPSPACE-hardness of the containment problem can be avoided. That is, there are fragments $\mathcal{F}$ of CRPQ whose containment problem is in PSPACE or even in lower classes. Which are these fragments? There are two natural systematic ways to define fragments of CRPQ, namely

1. as discussed before, by restricting the "shape" of the query, as in the underlying multigraph when regular expressions are abstracted away; or

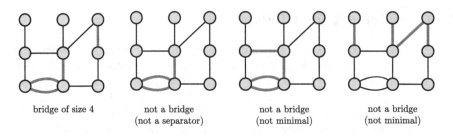

bridge of size 4	not a bridge	not a bridge	not a bridge
	(not a separator)	(not minimal)	(not minimal)

Fig. 3. Examples and non-examples of bridges.

2. by restricting the class of regular expressions that may occur in the queries RPQ atoms.

1. Restricting the Shape. Here we ask the same question as we did for the evaluation problem: given a class of multigraphs $\mathcal{C}$, is CONT-CRPQ($\mathcal{C}$) tractable? Of course here 'tractable' cannot be any better than PSPACE, since it is the complexity of CONT-RPQ, corresponding to the graph having two vertices and one edge. It turns out that, just as in the case for EVAL-CRPQ($\mathcal{C}$), one can characterize the classes of graphs $\mathcal{C}$ under which CONT-CRPQ($\mathcal{C}$) is in PSPACE. However, the graph measure is not treewidth but *bridgewidth*, which we define next.

A **bridge** of a (multi)graph is a minimal set of edges (in the sense of inclusion) whose removal increases the number of connected components (see Fig. 3 for some examples). The **bridge-width** of a graph is the maximum size of a bridge therein. Bridge-width is more restrictive than treewidth, in the sense that if a graph has bridge-width at most k then it also has treewidth at most k, but the converse does not necessarily hold. Let us define a class $\mathcal{C}$ of graphs to be **non-trivial** if it contains at least one graph with at least one edge; and let us call it **bridge-tame** if either $\mathcal{C}$ has bounded bridge-width or there is a polynomial time function $f : \mathbb{N} \to \mathcal{C}$ such that $f(n)$ has bridgewidth $\geq n$ for every n.

Theorem 8 ([25]). *For every non-trivial bridge-tame class $\mathcal{C}$ of graphs,*

- *if $\mathcal{C}$ has bounded bridge-width, then the containment problem for CRPQ($\mathcal{C}$) is PSPACE-complete;*
- *otherwise, the containment problem for CRPQ($\mathcal{C}$) is EXPSPACE-complete.*

2. Restricting the Regular Expressions. As we have remarked before, the lower bound construction of Theorems 7 and 8 make use of CRPQ which have a simple and regular shape (if seen as the underlying graph) but contain rather involved regular expressions, which do not correspond to CRPQ how they typically occur in practice. In fact, a large majority of regular expressions of queries used in practice are of a very simple form [16,17]. This motivates the study of CRPQ containment on fragments having commonly used kinds of regular expressions. The goal here is to identify restricted fragments of CRPQ that are both common in practice and have a reasonable complexity for query containment.

For a class of regular expressions $\mathcal{L}$, let CRPQ($\mathcal{L}$) be the set of CRPQ whose every RPQ atom uses an expression from $\mathcal{L}$.

Let $\mathcal{L}_s$ be the set of regular expressions of the form 's' for each symbol s of the finite alphabet. Let $\mathcal{L}_S$ be the set of expressions of the form '$a_1 + \cdots + a_n$' for $a_1, \ldots, a_n \in \mathbb{A}$ (*i.e.*, it corresponds to unions of $\mathcal{L}_s$). Finally, for $\alpha \in \{s, S\}$, let $\mathcal{L}_{\alpha^*}$ be the set of regular expressions of the form 'r^*' where $r \in \mathcal{L}_\alpha$. We next write $\mathcal{L}_{\alpha,\beta}$ as shorthand for $\mathcal{L}_\alpha \cup \mathcal{L}_\beta$.[4] Following this notation, observe that CRPQ($\mathcal{L}_s$) corresponds to the class of CQ (on graph databases), and that CRPQ($\mathcal{L}_S$) is contained in UCQ in terms of expressive power.

Theorem 9 ([23, 26]).

1. *The containment problem for CRPQ($\mathcal{L}_{S,S^*}$) and CRPQ($\mathcal{L}_{s,S^*}$) are* EXPSPACE-*complete.*
2. *The containment problem for CRPQ($\mathcal{L}_{s,s^*}$), CRPQ($\mathcal{L}_{S,s^*}$), and for CRPQ($\mathcal{L}_S$) are all Π_2^p-complete.*

Observe that CRPQ($\mathcal{L}$) is closed under concatenation in the following sense: Let $\mathcal{L}^{conc}$ be the closure under concatenation[5] of $\mathcal{L}$, then CRPQ($\mathcal{L}$) and CRPQ($\mathcal{L}^{conc}$) are equi-expressive (and there is a linear time translation from one to the other). This means that, for example, the Π_2^p upper bound for CRPQ($\mathcal{L}_{S,s^*}$) also holds for CRPQ having concatenations of expressions of $\mathcal{L}_{S,s^*}$ in the RPQ atoms, like $x \xrightarrow{(a+b) \cdot b^* \cdot (b+c)} y$. Notice also that, in light of the previous Theorem 8, the EXPSPACE lower bound of Theorem 9 uses —necessarily— queries of arbitrarily large bridge-width.

5 Boundedness of CRPQ

Boundedness is another important static analysis task of queries with a fixed-point feature. At an intuitive level, a query Q in any such logic is *bounded* if its fixed-point depth, *i.e.*, the number of iterations that are needed to evaluate Q on a database D, is bounded (and thus it is independent of the database D). In databases and knowledge representation, boundedness is regarded as an interesting theoretical phenomenon with relevant practical implications [14, 35]. In fact, while several applications in these areas require the use of recursive features, actual real-world systems are either not designed or not optimized to cope with the computational demands that such features impose. Bounded formulas, in turn, can be reformulated in 'non-recursive' logics, such as first-order logic, or even as a union of conjunctive queries (UCQ) when Q itself is positive. Since UCQs form the core of most systems for data management and ontological query answering, it is relevant to understand when a query can be equivalently translated to a UCQ as an optimization task. It has also been experimentally verified in some contexts that recursive features encountered in practice are often

---

[4] The choice of the fragments $\mathcal{L}_s$, $\mathcal{L}_S$, $\mathcal{L}_{s^*}$, and $\mathcal{L}_{S^*}$ is based on recent studies on SPARQL queries on Wikidata and DBpedia [13, 16, 17].

[5] That is, $\mathcal{L}^{conc} = \{s_1 \cdots s_n : \text{ for } n \in \mathbb{N} \text{ and } s_1, \ldots, s_n \text{ expressions from } \mathcal{L}\}$.

used in a somewhat 'harmless' way, and that many of such queries are in fact bounded [33]. We say that a CRPQ query is **bounded** if it is equivalent to some UCQ.

Example 6. Consider the following three Boolean CRPQ Q_1, Q_2, Q_3 over the alphabet $\mathbb{A} = \{a, b, c, d\}$ such that

$$Q_1 = (x \xrightarrow{L_b} y \wedge x \xrightarrow{L_{b,d}} y),$$
$$Q_2 = (x \xrightarrow{L_d} y \wedge x \xrightarrow{L_{b,d}} y),$$
$$Q_3 = (x \xrightarrow{L_b + L_d} y \wedge x \xrightarrow{L_{b,d}} y),$$

where $L_b = a^+b^+c$, $L_d = ad^+c^+$, and $L_{b,d} = a^+(b+d)c^+$. As it turns out, Q_1 and Q_2 are unbounded. However, Q_3 is bounded, and in particular, it is equivalent to the UCQ $\phi_1 \vee \phi_2$, where

$$\phi_1 = \exists x_0, x_1, x_2, x_3 \, (x_0 \xrightarrow{a} x_1) \wedge (x_1 \xrightarrow{b} x_2) \wedge (x_2 \xrightarrow{c} x_3)$$
$$\phi_2 = \exists x_0, x_1, x_2, x_3 \, (x_0 \xrightarrow{a} x_1) \wedge (x_1 \xrightarrow{d} x_2) \wedge (x_2 \xrightarrow{c} x_3).$$

◁

PROBLEM Boundedness problem for a class $\mathcal{Q}$ of (graph database) queries (BOUND-$\mathcal{Q}$)
GIVEN $Q \in \mathcal{Q}$
QUESTION Is there a UCQ Q' such that $Q \equiv Q'$?

For an RPQ $Q(x, y) = x \xrightarrow{L} y$ it is easy to see that the boundedness problem is really the finiteness problem of L: Q is bounded if, and only if, L is finite (*i.e.*, an NL-complete problem). However, if some of the variables are existentially quantified the problem is not as trivial. For example, a CRPQ of the form $\exists y \, x \xrightarrow{L} y$ is bounded if, and only if, the language

$$L_{prefix} = \{w \in L : \text{there is no proper prefix of } w \text{ in } L\}$$

is finite [8]. Likewise a CRPQ of the form $\exists x, y \, x \xrightarrow{L} y$ is bounded iff

$$L_{factor} = \{w \in L : \text{there is no proper factor of } w \text{ in } L\}$$

is finite. Both these problems are already PSPACE-complete [8]. For general CRPQ it turns out that the problem is related to the boundedness problem for an extension of finite automata which associate to each word in the language a natural number or 'cost', called *Distance Automata* [34] (*a.k.a.* weighted automata over the (min, +)-semiring [24], min-automata [15], or $\{\epsilon, ic\}$-B-automata [20]). The resulting complexity for the boundedness problem for CRPQ is, just as for the containment problem, EXPSPACE-complete.

Theorem 10 ([8]). BOUND-*CRPQ is* EXPSPACE-*complete. If a CRPQ is bounded, then it is equivalent to a UCQ of triple exponential size; and this bound is optimal.*

Contrary to the containment problem, very little is known when restricting either the shape or the languages of the CRPQ with regards to the boundedness problem.

6 Semantic Membership for CRPQ

As we have seen before, natural classes of CRPQ with tractable evaluation arise from considering bounded treewidth classes $\mathcal{C}$. However, this is a syntactic property, which begs the following question for any fixed class $\mathcal{C}$ of bounded treewidth: given a CRPQ Q, is it equivalent to some query Q' from CRPQ($\mathcal{C}$)? If so, we can replace the costly query Q with Q', or adapt the strategy for the (polynomial) evaluation of Q' to Q. The idea behind this optimization task is —as for boundedness— that the time needed to compute Q' may be comparatively small to the gain of having a polynomial time algorithm for the evaluation problem. Let $\mathcal{T}_k$ be the set of all multigraphs of treewidth at most k. We can then consider the following family of decision problems.

PROBLEM Treewidth-k semantic membership (MEM-tw_k)
GIVEN a CRPQ Q
QUESTION Is there a query Q' in CRPQ($\mathcal{T}_k$) such that $Q \equiv Q'$?

For the case where the input query turns out to be a Conjunctive Query his is a studied problem which is decidable, NP-complete [22] (basically, it reduces to testing treewidth of the *core* of a graph). However, for CRPQ this problem turns out to be more challenging. It has been shown to be decidable only for $k = 1$, that is, for 'trees', where trees should be understood as the class of multigraphs whose every simple cycle is of length 1 (*i.e.*, a self-loop) or 2 (*i.e.*, a cycle between a parent and a child).

Theorem 11 ([11]). MEM-tw_1 *is decidable,* EXPSPACE-*complete.*

It is however unknown if MEM-tw_k is decidable for any other k.

7 Extending CRPQ with Union and Two-Wayness

Two-wayness. Observe that semantics of RPQ and CRPQ are based on the notion of directed path. This means that the query cannot "freely" move around the graph edges, but it has to comply with the direction of edges. Remark that not even the reachability query "x and y belong to the same connected component in the underlying undirected graph" is expressible as a CRPQ. A standard extension of CRPQ and RPQ that palliates this lack of expressive power, consists in adding the ability to navigate the graph database with inverse relations, and it is known as **C2RPQ** and **2RPQ**, respectively (2 for "two-way navigation").

For any alphabet $\mathbb{A}$, let us define the alphabet $\mathbb{A}^{\pm} := \mathbb{A} \,\dot\cup\, \mathbb{A}^{-1}$ that extends $\mathbb{A}$ with the set $\mathbb{A}^{-1} := \{a^{-1} \mid a \in \mathbb{A}\}$ of "inverse" symbols. For a graph database G over $\mathbb{A}$, let $G^{\pm}$ be the result of adding an edge (v, a^{-1}, v') for every $(v', a, v) \in E$.

Now the regular expressions of a (C)2RPQ are defined over the extended alphabet $\mathbb{A}^{\pm}$. The semantics of a 2RPQ is extended as expected: a pair (v, v') of vertices of G satisfies a 2RPQ $x \xrightarrow{L} y$ if there is a path π in $G^{\pm}$ from v to v' such that $label(\pi) \in L$ (remember, now L is a regular subset of $(\mathbb{A}^{\pm})^*$). The semantics of C2RPQ follows the same definition based on 2RPQ.

Example 7. Consider the following 2RPQ

$$Q(x, y) = x \xrightarrow{(\text{supervise} \,\cdot\, \text{supervise}^{-1})^*} y.$$

On a graph database as the one of Example 1, Q returns pairs of people related by a "co-supervision" chain.   ◁

Union. A CRPQ, contrary to a CQ, has some restricted built-in union by the simple fact that regular languages are closed under union. However, the general structure of the query is fixed.

Example 8. Consider the following two Boolean CQs

$$Q_1 = x \xrightarrow{a} y$$
$$Q_2 = x \xrightarrow{b} x$$

It can be shown that there is no CRPQ expressing $Q_1 \vee Q_2$.   ◁

As for Conjunctive Query, it is a rather standard extension to add the possibility to have finite unions of queries, and it is known as **UCRPQ**. A UCRPQ is thus a query of the form $Q = Q_1 \vee \cdots \vee Q_n$, where every Q_i has the same set of free variables. We then define $\bar{x} \in Q(G)$ for a graph database G if $\bar{x} \in Q_i(G)$ for some i.

Finally, the extension including both possibilities of having two-way navigation as well as unions is denoted by **UC2RPQ**, and its semantics is as expected.

As it turns out, most known results extend to UC2RPQ in a seamless way: it just turns out that upper bound techniques (involving invariably some classes of automata) can extend to handle UC2RPQ (involving classes of *two-way* automata). In particular, the results on the containment and evaluation problems of Theorems 1, 3, 6, 7, 8, 10, and 11 extend to UC2RPQ while preserving the stated upper bounds.

8   Extending CRPQ with Path Comparison

Another studied extension of CRPQ conveys the ability to compare paths for certain relations on the labels. In comparison, one can think of CRPQ as testing for unary relations.

Example 9. Observe that the query Q_1 from Example 2 could be equivalently written as

$$Q_1(x) \;=\; \exists y \, \exists \pi_1, \pi_2 \; x \xrightarrow{\pi_1} y \wedge x \xrightarrow{\pi_2} y \wedge$$
$$label(\pi_1) \in L_{a^*b} \wedge label(\pi_2) \in L_{(a+b)^*c}$$

Where $L_\star$ is the language given by the expression $\star$. With this notation in mind, consider the following query

$$Q_2(x) \;=\; \exists y \, \exists \pi_1, \pi_2 \; x \xrightarrow{\pi_1} y \wedge y \xrightarrow{\pi_2} x \wedge$$
$$(label(\pi_1), label(\pi_2)) \in R$$

where $R \subseteq \mathbb{A}^* \times \mathbb{A}^*$ is now a *word relation* such as, for example, the equality relation $R = \{(u, v) \in \mathbb{A}^* \times \mathbb{A}^* : u = v\}$. Such query Q_2 would then output all vertices v having one cycling path with label $w\,w$ for some $w \in \mathbb{A}^*$.   ◁

Indeed, some scenarios require the ability to "compare paths", *i.e.*, to relate the words given by the labels corresponding to the existentially quantified paths in CRPQ. For example, when handling biological sequences, querying paths constrained under some *similarity* measure between paths is of great importance. See [9] for a more detailed discussion on the applicability of these features. The extension of CRPQ with non-monadic word relations gives rise to several expressive extensions which have been studied lately [3,4,7,9,10,43]. For a class of finite word relations $\mathcal{K}$ one can consider "CRPQ+$\mathcal{K}$", the result of extending CRPQ with testing of $\mathcal{K}$ relations on path labels.

Concretely, for any class $\mathcal{K}$ of finite word relations, the query language of **conjunctive regular path query with $\mathcal{K}$-relations (CRPQ+$\mathcal{K}$)** is a pair $(Q, \mathcal{R})$ where $\mathcal{R} \subseteq \mathcal{K}$ is a finite set of relations, each $R \in \mathcal{R}$ having arity $arity(R) \geq 1$. A CRPQ+$\mathcal{K}$ query Q, possibly having some free variables $\bar{x}$, is a query of the form

$$Q(\bar{x}) = \exists \bar{y} \, \exists \bar{\pi} \; \gamma(\bar{x}\bar{y}\bar{\pi}) \wedge \rho(\bar{\pi}). \tag{2}$$

It is a query over two sorts of (infinite, disjoint) sets of variables: **node variables** (denoting nodes of the graph database) and **path variables** (denoting paths). In (2), $\bar{x}$, $\bar{y}$ span over node variables and $\bar{\pi}$ over path variables. The idea is that γ tells how node variables are connected through path variables, while ρ describes the properties and relations between path variables in terms of the regular languages and relations of $\mathcal{R}$. Concretely, the subformula $\gamma(\bar{x}\bar{y}\bar{\pi})$, which we may call the **reachability subquery**, is a finite conjunction of **reachability atoms** of the form $z \xrightarrow{\pi} z'$, where z, z' are from $\bar{x}\bar{y}$ and π is from $\bar{\pi}$, with the restriction that every path variable π from $\bar{\pi}$ appears in exactly one reachability atom. That is, node variables may repeat in γ, but path variables may not. Let us call the subformula $\rho(\bar{\pi})$ the **relation subquery**, which is a finite conjunction of atoms of the form $R(\pi_1, \ldots, \pi_r)$, were $R \in \mathcal{R}$, $r = arity(R)$ and $\pi_1, \ldots, \pi_r$ are pairwise distinct path variables from $\bar{\pi}$.

For the classes of relations $\mathcal{K}$ we will consider here, we can think of each relation $R \in \mathcal{K}$ of arity $k \geq 1$ over an alphabet $\mathbb{A}$ as being described by an NFA over the alphabet of k-tuples $(\mathbb{A} \,\dot\cup\, \{\bot\})^k$. The underlying idea is that a word $w \in ((\mathbb{A} \,\dot\cup\, \{\bot\})^k)^*$ describes the k-tuple $(w_1, \ldots, w_k) \in (\mathbb{A}^*)^k$, where each w_i is obtained from w by (1) projecting onto the i-th component and (2) replacing each $\bot$ with the empty word ϵ. Thus, for example $(a, \bot, \bot)(b, b, \bot)(\bot, \bot, a)(c, \bot, \bot)$ describes (abc, b, a). In this way, any such NFA $\mathcal{A}$ denotes the k-ary relation R consisting of all tuples described by the words in the language of $\mathcal{A}$. Among the most basic classes of finite word relations are the classes of Recognizable, Synchronous (*a.k.a.* Automatic or Regular), and Rational relations [12]. The class of **Rational relations** is the set of all relations which are recognized by such automata, and it includes relations such as factor or subsequence. **Synchronous relations** are those that can be recognized by automata whose every word satisfies that, for every $i \leq k$, if a position has a $\bot$-symbol in its i-th component, then the next position (if it exists) must also have $\bot$ in its i-th component. For example, prefix or equal-length are synchronous relations. Finally, **Recognizable relations** are equivalent to finite unions of products of regular languages, *i.e.*, relations of the form $R = \bigcup_{i \in I} L_{i,1} \times \cdots \times L_{i,k}$ for a finite I, where the $L_{i,j}$'s are all regular languages over $\mathbb{A}$. They can also be defined in terms of NFA over $(\mathbb{A} \,\dot\cup\, \{\bot\})^k$ restricted to only accepting words such that: (1) no position contains more than one symbol from $\mathbb{A}$ and (2) every component projects onto a word from $\bot^* \cdot \mathbb{A}^* \cdot \bot^*$. An instance of a recognizable relation is $\{(u, v) : u, v \text{ start with the same letter}\}$ or $\{(u, v) : |u| + |v| = 3 \bmod 7\}$. These three classes form a proper hierarchy: Recognizable $\subsetneq$ Synchronous $\subsetneq$ Rational. Observe that any of the three classes of word relations contains the class of regular languages as (unary) relations. We refer the reader to [12] for more details on these classes.

Given a graph database $D = (V, E)$ over an alphabet $\mathbb{A}$ and an assignment f_n of $\bar{x}\bar{y}$ to V and an assignment f_p of $\bar{\pi}$ to paths of D, we say that (f_n, f_p) is a **satisfying assignment** if

1. for every reachability atom $z \xrightarrow{\pi} z'$ of γ, $f_p(\pi)$ is a directed path from $f_n(z)$ to $f_n(z')$ in D; and
2. for every r-ary atom $R(\pi_1, \ldots, \pi_r)$ of ρ the tuple

$$(label(f_p(\pi_1)), \ldots, label(f_p(\pi_r))) \in (\mathbb{A}^*)^r$$

is in the relation $R \in \mathcal{R}$.

Assuming $\bar{x} = (x_1, \ldots, x_\ell)$, the answers $Q(D)$ of the CRPQ+$\mathcal{K}$ query Q to the database D is the set of all $(f_n(x_1), \ldots, f_n(x_\ell)) \in V^\ell$ for every satisfying assignment (f_n, f_p).

It is plain to see that CRPQ+Recognizable is not more expressive than UCRPQ. On the other hand, the evaluation of CRPQ+Rational queries is undecidable, even for very simple rational relations. On the contrary, CRPQ+Synchronous seems to enjoy a good tradeoff of complexity and expressive power. This is partly because Synchronous relations constitute a very robust class, closed under Boolean operations and enjoying most of the decidability and algorithmic properties inherited from regular languages. CRPQ+Synchronous, commonly known as **ECRPQ** ('Extended' CRPQ).

Theorem 12.

- *(Folklore) CRPQ+Recognizable is contained in UCRPQ in terms of expressive power.*
- *[9]* EVAL-*CRPQ+Recognizable* and EVAL-*CRPQ+Synchronous* *(a.k.a. ECRPQ) are both* PSPACE-*complete [resp.* NL-*complete] in combined [resp. data] complexity.*
- *[9]* EVAL-*CRPQ+Rational is undecidable.*
- *[7]* EVAL-*CRPQ+(Synchronous* ∪ *{R}) is undecidable, for any R ∈ {suffix, factor}; EVAL-CRPQ+(Synchronous* ∪ *{subsequence}) is decidable and non-multiply-recursive hard.[6]*
- *[10]* EVAL-*CRPQ+{factor} is* PSPACE-*complete (both in data and combined complexity);* EVAL-*CRPQ+{subsequence} is* NEXPTIME-*complete [resp.* NP-*complete] in combined [resp. data] complexity.*

Observe that the data complexity for the evaluation of ECRPQ queries is the same as for CRPQ (*i.e.*, NL), but the combined complexity jumps from NP to PSPACE. In the same spirit as done for CRPQ in Theorem 3, there exists a characterization of the underlying structures $\mathcal{C}$ for which EVAL-ECRPQ($\mathcal{C}$) has better complexity. While the underlying structure of a CRPQ is the result of abstracting away its *languages*, the underlying structure of an ECRPQ is the result of abstracting away its *relations*. In contrast to the evaluation problem for CRPQ, the complexity of EVAL-ECRPQ($\mathcal{C}$) can be, depending on $\mathcal{C}$, either PTIME, NP, or PSPACE in combined complexity, and either XNL, W[1], or FPT in parameterized complexity [27]. Further, the FPT and PTIME cases do not coincide.

While the evaluation problem for ECRPQ remains decidable, the containment and equivalence problems turn out to be (roubstly) undecidable.

Theorem 13. CONT-*ECRPQ is undecidable [9]. Further, undecidability holds even for the fragment CRPQ+($\mathcal{RL}$ ∪ {eq-len}), where eq-len is the equal length binary relation and $\mathcal{RL}$ is the class of regular languages (seen as unary relations). In fact, this undecidability results even holds when one of the two inputs is a plain CRPQ (which is the same as CRPQ + $\mathcal{RL}$) [30].*

Other Extensions

A different extension to CRPQ with an expanded ability for path relational querying, consists in having "xregex" regular expressions with string variables (*a.k.a.* backreferences) [43], which is incomparable, in terms of expressive power, to ECRPQ.

As we remarked before, these graph query languages we have covered operate on an abstraction over a finite alphabet $\mathbb{A}$ of graph databases. However,

---

[6] In particular, this means that the time or space required by any algorithm deciding EVAL-CRPQ+(Synchronous ∪ {*subsequence*}) grows faster than the Ackermann function.

graph databases carry so called "data values" in the nodes and/or edges, that is, values with concrete domains, such as strings or numbers, and practical query languages can of course make tests on these data values, not only for equality but also using some domain-specific functions and relations. There have been proposals for querying mechanisms combining the ability to *test for data values* and *query the topology* [36]. In particular, one can explore different forms of querying paths while constraining the way data in paths changes. However, the theory of querying "data graphs" (*i.e.*, graph databases carrying elements from infinite domains) remains insofar a largely unexplored terrain.

The graph pattern languages here are based on a very simple form of recursion, namely applying regular expressions on paths. Another possible extension is by allowing a more complex form of recursion, such as nested regular expressions or Datalog-like rules, which increases the expressive power while often preserving complexities of CRPQ (see, *e.g.*, [41] and references therein).

9   Conclusions

We have explored some fundamental ways of querying graph databases via the concept of 'paths' on a simple abstraction of graph databases, that of edge-labelled graphs. One can draw a parallel between formalisms of CRPQ and its extensions and Conjunctive Queries, in the sense that these correspond to the most basic form of querying, using existentially quantified "patterns", by means of languages closed under homomorphisms. However, in this scenario, one deals with two-sorted languages dealing with *nodes* and *paths*, inheriting from the theory of words (or word relations) and of testing patterns (*i.e.*, Conjunctive Queries).

References

1. Angles, R., Arenas, M., Barceló, P., Hogan, A., Reutter, J.L., Vrgoc, D.: Foundations of modern query languages for graph databases. ACM Comput. Surv. **50**(5), 68:1-68:40 (2017). https://doi.org/10.1145/3104031
2. Angles, R., Gutiérrez, C.: Survey of graph database models. ACM Comput. Surv. **40**(1), 1:1-1:39 (2008). https://doi.org/10.1145/1322432.1322433
3. Anyanwu, K., Maduko, A., Sheth, A.P.: SPARQ2L: towards support for subgraph extraction queries in RDF databases. In: Williamson, C.L., Zurko, M.E., Patel-Schneider, P.F., Shenoy, P.J. (eds.) Proceedings of the 16th International Conference on World Wide Web, WWW 2007, Banff, Alberta, Canada, May 8–12, 2007, pp. 797–806. ACM (2007). https://doi.org/10.1145/1242572.1242680
4. Anyanwu, K., Sheth, A.P.: ρ-queries: enabling querying for semantic associations on the semantic web. In: Hencsey, G., White, B., Chen, Y.R., Kovács, L., Lawrence, S. (eds.) Proceedings of the Twelfth International World Wide Web Conference, WWW 2003, Budapest, Hungary, May 20–24, 2003, pp. 690–699. ACM (2003). https://doi.org/10.1145/775152.775249
5. Bagan, G., Bonifati, A., Groz, B.: A trichotomy for regular simple path queries on graphs. J. Comput. Syst. Sci. **108**, 29–48 (2020). https://doi.org/10.1016/j.jcss.2019.08.006

6. Barceló, P.: Querying graph databases. In: ACM Symposium on Principles of Database Systems (PODS), pp. 175–188. ACM (2013)
7. Barceló, P., Figueira, D., Libkin, L.: Graph logics with rational relations. Logic. Methods. Comput. Sci. (LMCS) **9**(3), 1 (2013). https://doi.org/10.2168/LMCS-9(3:1)2013
8. Barceló, P., Figueira, D., Romero, M.: Boundedness of conjunctive regular path queries. In: International Colloquium on Automata, Languages and Programming (ICALP). Leibniz International Proceedings in Informatics (LIPIcs), vol. 132, pp. 104:1–104:15. Leibniz-Zentrum für Informatik (2019). https://doi.org/10.4230/LIPIcs.ICALP.2019.104
9. Barceló, P., Libkin, L., Lin, A.W., Wood, P.T.: Expressive languages for path queries over graph-structured data. ACM Trans. Database Syst. (TODS) **37**(4), 31 (2012). https://doi.org/10.1145/2389241.2389250
10. Barceló, P., Muñoz, P.: Graph logics with rational relations: The role of word combinatorics. ACM Trans. Comput. Log. **18**(2), 10:1–10:41 (2017). https://doi.org/10.1145/3070822
11. Barceló, P., Romero, M., Vardi, M.Y.: Semantic acyclicity on graph databases. SIAM J. Comput. **45**(4), 1339–1376 (2016) https://doi.org/10.1137/15M1034714
12. Berstel, J.: Transductions and context-free languages, Teubner Studienbücher : Informatik, vol. 38. Teubner (1979). https://www.worldcat.org/oclc/06364613
13. Bielefeldt, A., Gonsior, J., Krötzsch, M.: Practical linked data access via SPARQL: the case of wikidata. In: Workshop on Linked Data on the Web (LDOW) (2018)
14. Bienvenu, M., Hansen, P., Lutz, C., Wolter, F.: First order-rewritability and containment of conjunctive queries in horn description logics. In: International Joint Conference on Artificial Intelligence (IJCAI), pp. 965–971 (2016)
15. Bojańczyk, M., Toruńczyk, S.: Deterministic automata and extensions of weak MSO. In: IARCS Annual Conference on Foundations of Software Technology and Theoretical Computer Science (FST&TCS). Leibniz International Proceedings in Informatics (LIPIcs), vol. 4, pp. 73–84. Leibniz-Zentrum für Informatik (2009). https://doi.org/10.4230/LIPIcs.FSTTCS.2009.2308
16. Bonifati, A., Martens, W., Timm, T.: Navigating the maze of Wikidata query logs. In: World Wide Web Conference (WWW), pp. 127–138 (2019)
17. Bonifati, A., Martens, W., Timm, T.: An analytical study of large SPARQL query logs. VLDB J. **29**, 655–679 (2019). https://doi.org/10.1007/s00778-019-00558-9
18. Calvanese, D., De Giacomo, G., Lenzerini, M., Vardi, M.Y.: Containment of conjunctive regular path queries with inverse. In: Principles of Knowledge Representation and Reasoning (KR), pp. 176–185 (2000)
19. Chandra, A.K., Merlin, P.M.: Optimal implementation of conjunctive queries in relational data bases. In: Hopcroft, J.E., Friedman, E.P., Harrison, M.A. (eds.) Proceedings of the 9th Annual ACM Symposium on Theory of Computing, May 4–6, 1977, Boulder, Colorado, USA, pp. 77–90. ACM (1977). https://doi.org/10.1145/800105.803397
20. Colcombet, T.: The theory of stabilisation monoids and regular cost functions. In: Albers, S., Marchetti-Spaccamela, A., Matias, Y., Nikoletseas, S., Thomas, W. (eds.) ICALP 2009. LNCS, vol. 5556, pp. 139–150. Springer, Heidelberg (2009). https://doi.org/10.1007/978-3-642-02930-1_12
21. Consens, M.P., Mendelzon, A.O.: Graphlog: a visual formalism for real life recursion. In: Rosenkrantz, D.J., Sagiv, Y. (eds.) Proceedings of the Ninth ACM SIGACT-SIGMOD-SIGART Symposium on Principles of Database Systems, April 2–4, 1990, Nashville, Tennessee, USA, pp. 404–416. ACM Press (1990). https://doi.org/10.1145/298514.298591

22. Dalmau, V., Kolaitis, P.G., Vardi, M.Y.: Constraint satisfaction, bounded treewidth, and finite-variable logics. In: Van Hentenryck, P. (ed.) CP 2002. LNCS, vol. 2470, pp. 310–326. Springer, Heidelberg (2002). https://doi.org/10.1007/3-540-46135-3_21

23. Deutsch, A., Tannen, V.: Optimization properties for classes of conjunctive regular path queries. In: Ghelli, G., Grahne, G. (eds.) DBPL 2001. LNCS, vol. 2397, pp. 21–39. Springer, Heidelberg (2002). https://doi.org/10.1007/3-540-46093-4_2

24. Droste, M., Kuich, W., Vogler, H.: Handbook of weighted automata. Springer Science & Business Media, Heidelberg (2009)

25. Figueira, D.: Containment of UC2RPQ: the hard and easy cases. In: International Conference on Database Theory (ICDT). Leibniz International Proceedings in Informatics (LIPIcs), Leibniz-Zentrum für Informatik (2020)

26. Figueira, D., Godbole, A., Krishna, S., Martens, W., Niewerth, M., Trautner, T.: Containment of simple conjunctive regular path queries. In: Principles of Knowledge Representation and Reasoning (KR) (2020). https://hal.archives-ouvertes.fr/hal-02505244

27. Figueira, D., Ramanathan, V.: When is the evaluation of Extended CRPQ tractable? (2021). https://hal.archives-ouvertes.fr/hal-03353483. (working paper or preprint)

28. Florescu, D., Levy, A., Suciu, D.: Query containment for conjunctive queries with regular expressions. In: ACM Symposium on Principles of Database Systems (PODS), pp. 139–148. ACM Press (1998). https://doi.org/10.1145/275487.275503

29. Flum, J., Grohe, M.: Parameterized Complexity Theory. Texts in Theoretical Computer Science. An EATCS Series, Springer, Heidelberg (2006). https://doi.org/10.1007/3-540-29953-X

30. Freydenberger, D.D., Schweikardt, N.: Expressiveness and static analysis of extended conjunctive regular path queries. J. Comput. Syst. Sci. **79**(6), 892–909 (2013) https://doi.org/10.1016/j.jcss.2013.01.008

31. Grohe, M., Schwentick, T., Segoufin, L.: When is the evaluation of conjunctive queries tractable? In: Vitter, J.S., Spirakis, P.G., Yannakakis, M. (eds.) Proceedings on 33rd Annual ACM Symposium on Theory of Computing, July 6–8, 2001, Heraklion, Crete, Greece. pp. 657–666. ACM (2001). https://doi.org/10.1145/380752.380867, https://doi.org/10.1145/380752.380867

32. Gyssens, M., Paredaens, J., Gucht, D.V.: A graph-oriented object database model. In: Rosenkrantz, D.J., Sagiv, Y. (eds.) Proceedings of the Ninth ACM SIGACT-SIGMOD-SIGART Symposium on Principles of Database Systems, April 2–4, 1990, Nashville, Tennessee, USA, pp. 417–424. ACM Press (1990). https://doi.org/10.1145/298514.298593

33. Hansen, P., Lutz, C., Seylan, I., Wolter, F.: Efficient query rewriting in the description logic EL and beyond. In: International Joint Conference on Artificial Intelligence (IJCAI), pp. 3034–3040 (2015)

34. Hashiguchi, K.: Limitedness theorem on finite automata with distance functions. J. Comput. Syst. Sci. (JCSS) **24**(2), 233–244 (1982)

35. Hillebrand, G.G., Kanellakis, P.C., Mairson, H.G., Vardi, M.Y.: Tools for datalog boundedness. In: ACM Symposium on Principles of Database Systems (PODS), pp. 1–12 (1991)

36. Libkin, L., Martens, W., Vrgoc, D.: Querying graphs with data. J. ACM **63**(2), 14:1–14:53 (2016). https://doi.org/10.1145/2850413

37. Malyshev, S., Krötzsch, M., González, L., Gonsior, J., Bielefeldt, A.: Getting the most out of Wikidata: semantic technology usage in Wikipedia's knowledge graph. In: International Semantic Web Conference (ISWC), pp. 376–394 (2018)

38. Manola, F., et al.: Rdf primer. W3C Recommend. **10**(1–107), 6 (2004)
39. Martens, W., Niewerth, M., Trautner, T.: A trichotomy for regular trail queries. In: Paul, C., Bläser, M. (eds.) 37th International Symposium on Theoretical Aspects of Computer Science, STACS 2020, March 10–13, 2020, Montpellier, France. LIPIcs, vol. 154, pp. 7:1–7:16. Schloss Dagstuhl - Leibniz-Zentrum für Informatik (2020). https://doi.org/10.4230/LIPIcs.STACS.2020.7, https://doi.org/10.4230/LIPIcs.STACS.2020.7
40. Mendelzon, A.O., Wood, P.T.: Finding regular simple paths in graph databases. SIAM J. Comput. **24**(6), 1235–1258 (1995) https://doi.org/10.1137/S009753979122370X
41. Reutter, J.L., Romero, M., Vardi, M.Y.: Regular queries on graph databases. Theory Comput. Syst. **61**(1), 31–83 (2017) https://doi.org/10.1007/s00224-016-9676-2
42. Romero, M., Barceló, P., Vardi, M.Y.: The homomorphism problem for regular graph patterns. In: Annual Symposium on Logic in Computer Science (LICS), pp. 1–12. IEEE Computer Society Press (2017). https://doi.org/10.1109/LICS.2017.8005106
43. Schmid, M.L.: Conjunctive regular path queries with string variables. In: Suciu, D., Tao, Y., Wei, Z. (eds.) Proceedings of the 39th ACM SIGMOD-SIGACT-SIGAI Symposium on Principles of Database Systems, PODS 2020, Portland, OR, USA, June 14–19, 2020, pp. 361–374. ACM (2020). https://doi.org/10.1145/3375395.3387663
44. Vardi, M.Y.: The complexity of relational query languages (extended abstract). In: Lewis, H.R., Simons, B.B., Burkhard, W.A., Landweber, L.H. (eds.) Proceedings of the 14th Annual ACM Symposium on Theory of Computing, May 5–7, 1982, San Francisco, California, USA, pp. 137–146. ACM (1982). https://doi.org/10.1145/800070.802186
45. Wood, P.T.: Query languages for graph databases. SIGMOD Rec. **41**(1), 50–60 (2012) https://doi.org/10.1145/2206869.2206879

On Combining Ontologies and Rules

Matthias Knorr$^{(\boxtimes)}$ (iD)

Universidade Nova de Lisboa, Caparica, Portugal
`mkn@fct.unl.pt`

Abstract. Ontology languages, based on Description Logics, and non-monotonic rule languages are two major formalisms for the representation of expressive knowledge and reasoning with it, that build on fundamentally different ideas and formal underpinnings. Within the Semantic Web initiative, driven by the World Wide Web Consortium, standardized languages for these formalisms have been developed that allow their usage in knowledge-intensive applications integrating increasing amounts of data on the Web. Often, such applications require the advantages of both these formalisms, but due to their inherent differences, the integration is a challenging task. In this course, we review the two formalisms and their characteristics and show different ways of achieving their integration. We also discuss an available tool based on one such integration with favorable properties, such as polynomial data complexity for query answering when standard inference is polynomial in the used ontology language.

1 Introduction

Though the central idea of Artificial Intelligence (AI) – creating autonomous agents that are able to think, act, and interact in a rational manner [69] – is far from being fullfilled, the results of the scientific advances of this major endavour already affect our daily lifes and our society. Such applications of AI technologies include, among others, clinical decision support systems in Healthcare, fraud dectection in financial transactions, (semi-)autonomous vehicles, crop and soil monitoring in agriculture, surveillance for border protection, but also in the form of personal assistants in cell phones. In these cases, technologies from different subareas of AI are applied, such as Learning, Vision, Reasoning, Planning, or Speech Recognition, but common to them is the requirement of efficiently processing and taking advantage of huge amounts of data, for example, for finding patterns in a data set, or for determining a conclusion based on the given data to aid making a decision.

Knowledge Representation and Reasoning (KRR) [14] in particular is an area in AI that aims at representing knowledge about the world or a domain of interest, commonly relying on logic-based formalisms, and that allows the usage of automated methods to reason with this knowledge and draw conclusions from it to aid in the beforementioned applications. Due to the structured formalization of knowledge, technologies based on KRR provide provenly correct

© Springer Nature Switzerland AG 2022
M. Šimkus and I. Varzinczak (Eds.): Reasoning Web 2021, LNCS 13100, pp. 22–58, 2022.
https://doi.org/10.1007/978-3-030-95481-9_2

conclusions and are commonly amenable to complement these with justifications or explanations for such derived results, which is important for accountabilility in applications where critical decisions need to be made, such as in Healthcare or in the financial sector. Yet, though KRR formalisms are responsible for many early success stories in the history of AI, namely in the form of expert systems, it has been pointed out that creating such formalizations often requires a major effort for each distinct such application [69].

As it turns out, a solution to the latter problem can be found in the area of the Semantic Web [39]. The central idea of the Semantic Web can be described as extending the World Wide Web, mainly targetted at human consumers, to one where data on the internet would be machine-processable, giving rise to advanced applications [12]. To achieve that goal, web pages are extended with machine-processable data written in standardized languages developed by the World Wide Web Consortium[1] (W3C) that allow the specification and identification of common terms in a uniformized way. In this context, the Linked Open Data initiative appeared leading to the publication of large interlinked datasets, utilizing these standardized languages, that are distributed over the Web. The resulting Linked Open Data Cloud[2] covers areas such as geography, government, life sciences, linguistics, media, scientific publications, and social networking, including data from, e.g., DBpedia [52], containing data extracted from Wikipedia, and prominent industry contributers such as the BBC, with accounted industry adopters, such as the New York Times Company or Facebook.

These developments facilitate the access and reuse of data and knowledge published on the Web in these standardized languages and thus also the creation of knowledge-intensive applications. The standards include the Resource Description Framework (RDF) [70] for describing information on directed, labelled and typed graphs, based on XML syntax; the Web Ontology Language (OWL) [40] for specifying shared conceptualizations and taxonomic knowledge; the Rule Interchange Format (RIF) [60] for expressing inferences structured in the form of rules, and SPARQL [38] for querying RDF knowledge. Among them, two are of particular interest for the representation of expressive knowledge, namely ontology languages and rule languages with different characteristics.

Ontology languages are widely used to represent and reason over formal conceptualizations of hierarchical, taxonomic knowledge. OWL in particular is founded on Description Logics (DLs) [5], which are commonly decidable fragments of first-order logic. Due to that, DLs are monotonic by nature, which means that acquiring new information does not invalidate previously drawn conclusions. Also, they apply the Open World Assumption (OWA), which means that no conclusions can be drawn merely based on the absence of some piece of information. They also allow us to reason over abstract relations between classes of objects, without the need to refer to concrete instances or objects, as well as to

---

[1] http://www.w3.org.
[2] https://lod-cloud.net/.

reason with unknown individuals, i.e., objects that are inferred to exist though they do not correspond to any known object in the represented knowledge. Different DLs are defined based on varying combinations of logical operators that are admitted to be used, which allows one to choose a concrete language that balances between the required expressiveness of representation of the language and the resulting complexity of reasoning with it. This is why, in addition to the language standard OWL 2, which is very expressive but also comes with a high worst-complexity of reasoning, so-called profiles [61] of OWL 2 have been defined, each with different application areas in mind, that limit the admitted operators, but allow for polynomial reasoning.

Rule languages. There is a large variety of rule languages, which in their essence can be described as expressing *IF - THEN* relationships. They are commonly divided into production rules, that can be viewed as describing conditional effects, and declarative rules that describe knowledge about the world. This is reflected in RIF in the sense that it is actually not a single language standard, but rather a number of different formats, so-called dialects, that can be integrated by means of RIF. Here, our interest resides on declarative languages to represent knowledge and reason with it, in that they represent inferences from premises to conclusions, commonly on the level of concrete instances. Among these languages, nonmonotonic rules as established originally in Logic Programming (LP) [10] are of particular interest. Such nonmonotonic rules employ the Closed World Assumption (CWA), i.e., if something cannot be derived to be true, it is assumed false. These conclusions may be revised in the presence of additional information, hence the term nonmonotonic. Nonmonotonic rules are thus well-suited for modelling default knowledge and exceptions, in the sense that commonly a certain conclusion can be drawn unless some exceptional characteristic prevents the conclusion, as well as integrity constraints that allow us to ensure that certain required specifications on the data are met.

Integration. Due to the different characteristics of the two formalisms, their integration is often necessary in a variety of applications (see e.g., [1,45,55,65, 74]). To illustrate these requirements, we present some examples of such use cases, starting with an example which we will revisit throughout the paper.

Example 1. Consider a customs service that needs to assess incoming cargo in a port for a variety of risk factors including terrorism, narcotics, food and consumer safety, pest infestation, tariff violations, and intellectual property rights. The assessment of the mentioned risks requires taking into consideration extensive knowledge about commodities, business entities, trade patterns, government policies and trade agreements. While some of this knowledge is external to the considered customs agency, such as the classification of commodities according to the international Harmonized Tariff System (HTS), or knowledge about international trade agreements, other parts are internal, such as policies that help determine which cargo to inspect (under which circumstances), as well as, for example, records on the history of prior inspections that allow one to identify importers with a history of good compliance with the regulations and those that require closer monitoring due to previous infractions.

In this setting, ontologies are a natural choice to model among others the international taxonomy of commodities, whereas rules are more suitable to model, e.g., the policies that determine which cargo to inspect. Notably, the distinct characteristics of the two formalisms are important to represent different parts of the desired knowledge. On the one hand, when reasoning about whether a certain importer requires close monitoring, the fact that we cannot find any previous infractions suffices to conclude by default that this importer is not a suspect. This aligns well with the CWA usually employed in nonmonotonic rules. On the other hand, the fact that we do not know whether a certain product is contained in a cargo container, should not allow us to conclude that it is not there. Here, the OWA of ontology languages is more suitable. At the same time, with ontology languages we are able to reason abstractly with commodities, for example listed in the manifest of a container without having to refer to concrete objects.

Another example can be found in clinical health care, where large ontologies, such as SNOMED CT,[3] are used for electronic health record systems and clinical decision support, but where nonmonotonic rules are required to express conditions such as dextrocardia, i.e., that the heart is exceptionally on the right side of the body. Also, when matching patient records with clinical trials criteria [65], modeling pharmacy data of patients using the CWA is cleary preferable (to avoid listing all the medications not taken), because usually it can be assumed that a patient is not under a specific medication unless explicitly known, whereas conclusions on medical conditions (or the absence of them) require explicit proof, e.g., based on test results. A further use-case can be found in the maintainance of a telecommunications inventory [45], where an ontology is used to represent the hierarchy of existing equipment (with a uniform terminology used for such equipment in different countries), and nonmonotonic rules are applied to detect different failures based on current sensor data.

However, the combination of ontologies and nonmonotonic rules is difficult due to the inherent differences between the underlying base formalisms and the way how decidability of reasoning is achieved in them (see, e.g., [27,28,62]). This raises several questions, such as:

- How to achieve such an integration given the inherent technical differences?
- What is a suitable topology, i.e., should both formalisms be on equal terms or should one's inferences only serve as input for the other (and not vice-versa)?
- How and when should Open or Closed World Assumption be applied?
- How to ensure decidability for the integration?
- Can we obtain efficient reasoning procedures?

In this lecture, we discuss how to answer these questions. For this purpose, after giving an overview on the two base formalisms, we review major approaches

---

[3] http://www.ihtsdo.org/snomed-ct/.

that have been introduced in the literature to tackle this problem, and we identify essential criteria along which these (and many other such) integrations have been introduced. These criteria may also help choose the most suitable appproach among them for a concrete application. We then also discuss an available tool based on one such integration with favorable properties along the established criteria, including efficient reasoning procedures.

The remainder of this document is structured as follows. We provide necessary notions on Description Logic ontologies and nonmonotonic rules in Sects. 2 and 3, respectively. We then discuss the principles along which integrations between the two formalisms have been developed in Sect. 4, and show, in Sect. 5 four such integrations in more detail. We proceed with presenting a tool for answering queries over such an integration in Sect. 6 and finish with concluding remarks in Sect. 7.

2   Description Logic Ontologies

Description Logics (DLs) are fragments of first-order logics for which reasoning is usually decidable. They are commonly used as expressive ontology languages, in particular as the formal underpinning for the Web Ontology Language (OWL) [40]. In this section, we provide an overview on DLs to facilitate the understanding of the remaining material and refer for more details and additional background to text books in related work [5,41].

On a general level, DLs allow us represent information about objects, called individuals, classes of objects, called concepts, that share common characteristics, and relations between objects as well as between classes of objects, called roles. For example, Bob is an individual, $Person$ a concept, and $hasSSN$ a role abbreviating "has social security number" allowing us to state that Bob is a person and that he has a specific social security number.

More formally, DLs are defined over disjoint countably infinite sets of concept names N_C, corresponding to classes of individuals, role names N_R, that express relationships, and individual names N_I, corresponding to objects, which, in terms of first-order logic, match unary and binary predicates, and constants, respectively.

Complex concepts (and complex roles) can be defined based on these sets and logical constructors (indicated here in DL syntax), which include the standard Boolean operators, i.e., negation ($\neg$), conjunction ($\sqcap$), and disjunction ($\sqcup$), universal and existential quantifiers ($\forall$ and $\exists$), as well as numeric restrictions on (binary) roles ($\leq mR$ and $\geq nR$ for numbers m, n and role R), and inverses of roles (R^- for role R).

For example, we can define concepts that represent persons that have a social security number ($Person \sqcap \exists hasSNN.\top$), or those whose heart is on the left- or on the right-hand-side of the body ($HeartLeft \sqcup HeartRight$), or those having at most one social security number ($\leq 1HasSNN.\top$) or the role "is the social security number of" as the inverse of the role of having a social security number ($hasSSN^-$).

Name	Syntax	Semantics
inverse role	R^-	$\{(x,y) \in \Delta^{\mathcal{I}} \times \Delta^{\mathcal{I}} \mid (y,x) \in R^{\mathcal{I}}\}$
universal role	U	$\Delta^{\mathcal{I}} \times \Delta^{\mathcal{I}}$
top	$\top$	$\Delta^{\mathcal{I}}$
bottom	$\bot$	$\emptyset$
negation	$\neg C$	$\Delta^{\mathcal{I}} \setminus C^{\mathcal{I}}$
conjunction	$C \sqcap D$	$C^{\mathcal{I}} \cap D^{\mathcal{I}}$
disjunction	$C \sqcup D$	$C^{\mathcal{I}} \cup D^{\mathcal{I}}$
nominals	$\{a\}$	$\{a^{\mathcal{I}}\}$
universal restriction	$\forall R.C$	$\{x \in \Delta^{\mathcal{I}} \mid (x,y) \in R^{\mathcal{I}} \text{ implies } y \in C^{\mathcal{I}}\}$
existential restriction	$\exists R.C$	$\{x \in \Delta^{\mathcal{I}} \mid \exists y \in \Delta^{\mathcal{I}}, (x,y) \in R^{\mathcal{I}} \wedge y \in C^{\mathcal{I}}\}$
Self concept	$\exists R.\mathsf{Self}$	$\{x \in \Delta^{\mathcal{I}} \mid (x,x) \in R^{\mathcal{I}}\}$
qualified number	$\leqslant nR.C$	$\{x \in \Delta^{\mathcal{I}} \mid \sharp\{(x,y) \in R^{\mathcal{I}} \text{ and } y \in C^{\mathcal{I}}\} \leq n\}$
restrictions	$\geqslant nR.C$	$\{x \in \Delta^{\mathcal{I}} \mid \sharp\{(x,y) \in R^{\mathcal{I}} \text{ and } y \in C^{\mathcal{I}}\} \geq n\}$

Fig. 1. Syntax and semantics of role and concept expressions in $\mathcal{SROIQ}$ for an interpretation $\mathcal{I}$ with domain $\Delta^{\mathcal{I}}$, where C, D are concepts, R a role, and $a \in \mathsf{N_I}$.

It should be noted that each DL, i.e., each fragment of first-order logic, can be distinguished by the admitted constructors, by means of which complex concepts and roles are inductively formed. An overview on the admitted constructors in the DL underlying the OWL 2 language, i.e., the description logic $\mathcal{SROIQ}$ [42], can be found in Fig. 1.

Based on complex concepts and roles, we can define an ontology containing axioms that specify a taxonomy/conceptualization of a domain of interest. Such axioms allow us to indicate that a concept is subsumed by another, i.e., corresponding to a subclass relation, or that two concepts are equivalent ($\equiv$) as a shortcut for subsumption in both directions, and likewise that some role is subsumed by (or equivalent to) another role, as well as assertions indicating that some individual is an instance of a class (or a pair of individuals an instance of a role), and even certain role characteristics, such as transitivity, reflexivity or symmetry of roles.

For example, the following set of axioms

$$Person \sqsubseteq HeartLeft \sqcup HeartRight \tag{1}$$
$$Person \sqsubseteq \exists hasSNN.\top \tag{2}$$
$$isSSNOf \equiv hasSSN^- \tag{3}$$
$$Person(Bob) \tag{4}$$

indicates that every person has the heart on the left or on the right-hand-side (1), every person has a social security number (2), the relation referring to the person corresponding to an SSN is the inverse of the relation indicating the SSN of a person (3), and that Bob is a person (4).

Formally, an *ontology* $\mathcal{O}$ is a set of axioms divided into three components, a TBox, an RBox, and an ABox. A *TBox* is a finite set of *concept inclusions* of the form $C \sqsubseteq D$ where C and D are both (complex) concepts. An *RBox* is a finite set of *role chains* of the form $R_1 \circ \ldots \circ R_n \sqsubseteq R$, including *role inclusions* with $n = 1$, where R_i and R are (complex) roles, and of *role assertions*, allowing us to express that a role is transitive (Tra), reflexive (Ref), irreflexive (Irr), symmetric (Sym) or asymmetric (Asy), and that two roles are disjoint (Dis). Finally, an *ABox* is a finite set of *assertions* of the form $C(a)$ or $R(a, b)$ for concepts C, roles R, and individuals a, b.

The semantics of DL ontologies, i.e., their meaning, is defined in terms of interpretations $\mathcal{I} = (\Delta^{\mathcal{I}}, \cdot^{\mathcal{I}})$ consisting of a non-empty set $\Delta^{\mathcal{I}}$, its domain, and a mapping $\cdot^{\mathcal{I}}$ that specifies how the basic language elements are to be interpreted. More concisely, such a mapping ensures that

- $a^{\mathcal{I}} \in \Delta^{\mathcal{I}}$ for each $a \in \mathsf{N_I}$;
- $A^{\mathcal{I}} \subseteq \Delta^{\mathcal{I}}$ for each $A \in \mathsf{N_C}$; and
- $R^{\mathcal{I}} \subseteq \Delta^{\mathcal{I}} \times \Delta^{\mathcal{I}}$ for each $R \in \mathsf{N_R}$.

That is, every individual is interpreted as one domain element, every concept as a subset of the domain, and every role as a set of pairs of domain elements. This mapping can be extended to arbitrary role and concept expressions as indicated in Fig. 1 for the constructors usable in $\mathcal{SROIQ}$.

The semantics of axioms α is based on a satisfaction relation w.r.t. an interpretation $\mathcal{I}$, denoted $\mathcal{I} \models \alpha$, which is presented in Fig. 2. In this case, we also say that $\mathcal{I}$ is a model of α, and likewise that $\mathcal{I}$ is a model of ontology $\mathcal{O}$ if $\mathcal{I}$ satisfies all axioms in $\mathcal{O}$. It should be noted that an alternative way to define the semantics of ontologies is obtained by a direct translation into first-order logic. For example, (2) can be translated into

$$\forall x.(Person(x) \rightarrow (\exists y.hasSNN(x, y)))$$

Note that the syntax of DLs does not explicitly mention the variables introduced in their first-order translation. They are left implicit, and, due to the restriction to unary and binary predicates (concepts and roles) and their particular structure, no ambiguity exists. We refer for the details of this translation to [5].

Based on the semantics in place, we can consider reasoning and what kinds of inferences can be drawn. For example, from

$$Person \sqsubseteq \exists has.SpinalColumn \tag{5}$$

$$\exists has.SpinalColumn \sqsubseteq Vertebrate \tag{6}$$

we can conclude that $Person \sqsubseteq Vertebrate$ holds and we can draw this conclusion without needing to know any particular individual person. In addition, together with (4), we can also conclude that Bob in particular is a vertebrate,

Axiom α	Condition for $\mathcal{I} \models \alpha$
$R_1 \circ \ldots \circ R_n \sqsubseteq R$	$R_1^{\mathcal{I}} \circ \ldots \circ R_n^{\mathcal{I}} \sqsubseteq R^{\mathcal{I}}$
$\mathsf{Tra}(\mathsf{R})$	if $R^{\mathcal{I}} \circ R^{\mathcal{I}} \subseteq R^{\mathcal{I}}$
$\mathsf{Ref}(\mathsf{R})$	$(x, x) \in R^{\mathcal{I}}$ for all $x \in \Delta^{\mathcal{I}}$
$\mathsf{Irr}(\mathsf{R})$	$(x, x) \notin R^{\mathcal{I}}$ for all $x \in \Delta^{\mathcal{I}}$
$\mathsf{Dis}(\mathsf{R}, \mathsf{S})$	if $(x, y) \in R^{\mathcal{I}}$, then $(x, y) \notin S^{\mathcal{I}}$ for all $x, y \Delta^{\mathcal{I}}$
$\mathsf{Sym}(\mathsf{R})$	if $(x, y) \in R^{\mathcal{I}}$, then $(y, x) \in R^{\mathcal{I}}$ for all $x, y \in \Delta^{\mathcal{I}}$
$\mathsf{Asy}(\mathsf{R})$	if $(x, y) \in R^{\mathcal{I}}$, then $(y, x) \notin R^{\mathcal{I}}$ for all $x, y \in \Delta^{\mathcal{I}}$
$C \sqsubseteq D$	$C^{\mathcal{I}} \subseteq D^{\mathcal{I}}$
$C(a)$	$a^{\mathcal{I}} \in C^{\mathcal{I}}$
$R(a, b)$	$(a^{\mathcal{I}}, b^{\mathcal{I}}) \in R^{\mathcal{I}}$
$\neg R(a, b)$	$(a^{\mathcal{I}}, b^{\mathcal{I}}) \notin R^{\mathcal{I}}$
$a \not\approx b$	$a^{\mathcal{I}} \neq b^{\mathcal{I}}$

Fig. 2. Semantics of $\mathcal{SROIQ}$ axioms for an interpretation $\mathcal{I}$ with domain $\Delta^{\mathcal{I}}$, where C, D are concepts, R, S are roles, and $a, b \in \mathsf{N_I}$.

and, by (2) and (4), that there is some object which is a social security number, although we do not know precisely which one.

In more general terms, a number of standard inference problems are defined for DLs. For example, we can test if a DL ontology $\mathcal{O}$ is *consistent* by determining if it has a model. We can also determine if a concept is *satisfiable* by finding a model $\mathcal{I}$ in which the interpretation of C is not empty ($C^{\mathcal{I}} \neq \emptyset$). Also, concept C *is subsumed by* concept D w.r.t. $\mathcal{O}$, i.e., C is a subclass of D, if $C^{\mathcal{I}} \subseteq D^{\mathcal{I}}$ holds for all models $\mathcal{I}$ of $\mathcal{O}$. Moreover, we can check whether an individual a is an *instance of a concept* C w.r.t. $\mathcal{O}$, if $a^{\mathcal{I}} \in C^{\mathcal{I}}$ holds for all models $\mathcal{I}$ of $\mathcal{O}$. Frequently, these reasoning tasks can be reduced to each other, e.g., C is subsumed by D iff $C \sqcap \neg D$ is unsatisfiable, which means that it suffices to consider only one of them. Also, based on these standard inference problems, advanced reasoning tasks are considered, such as *classification* which requires the computation of the subsumptions between all concept names in the ontology, or *instance retrieval* which determines all individuals that are true for a given concept.

Decidability is achieved by carefully restricting the admitted constructors such that models satisfy certain desirable properties. This is necessary, as the domain of an interpretation $\Delta^{\mathcal{I}}$ may be infinite. Note that DLs only allow the usage of unary and binary predicates which together with the admitted constructors ensures that basic DLs satisfy the tree-model property, i.e., domain elements in a model are related in a tree-shaped manner, which avoids cycles, and thus ensures decidability. For more expressive DLs such as $\mathcal{SROIQ}$ this does not suffice. In fact, if we were allowed to use the constructors in Figs. 1 and 2 without additional restrictions, reasoning in $\mathcal{SROIQ}$ would be undecidable. Instead, role hierarchies, i.e., essentially axioms involving role inclusions, are restricted to so-called regular ones, which intuitively limits the usage of certain roles in, e.g., qualified number restrictions and role chains, so that a strict order can be established in between the roles. We refer for the details to [42].

While the applied restrictions ensure that $\mathcal{SROIQ}$ is decidable, this DL underlying the W3C standard OWL 2 is still very general and highly expressive, which in turn implies that reasoning with it is highly complex, in fact, it is N2ExpTime-complete [46]. But general purpose DL reasoners exist for this highly expressive language such as FacT++ [79], Pellet [71], Racer [37] or Konclude [75]. Still, one of the central themes in DLs is the balance between expressiveness and the computational complexity for the required reasoning tasks. This is why the profiles OWL 2 EL, OWL 2 QL and OWL 2 RL have been defined [61], i.e., considerable restrictions of the admitted language of OWL 2, for which reasoning is tractable. Here, we briefly overview these languages.

The DL underlying OWL 2 EL is $\mathcal{EL}^{++}$ [6]. Often, $\mathcal{EL}^+_\bot$, a large fragment of it is used that only allows conjunctions and existential restrictions of concepts, hierarchies of roles, and disjoint concepts. $\mathcal{EL}^+_\bot$ is tailored towards reasoning with large conceptual models, i.e., large TBoxes, and used in particular in large biomedical ontologies, and specifically tailored, highly efficient reasoners such as ELK [47] exist.

$DL\text{-}Lite_R$, one of the languages of the $DL\text{-}Lite$ family [4] which underlies OWL 2 QL, admits in addition inverse roles and even disjoint roles, but, in exchange, it only allows for simple role hierarchies, no conjunction (on the left-hand side of axioms), and it imposes limitations on the usage of existential restrictions in particular on the left-hand side of inclusion axioms. The focus of this profile is on answering queries over large quantities of data, and combining relational database technology in the context of ontology-based data access [81].

Finally, OWL 2 RL, which builds on Description Logic Programs [36], is defined in a way that avoids inferring the existence of unknown individuals (e.g., a social security number though we do not know which) as well as nondeterministic knowledge, for the sake of efficient reasoning. OWL 2 RL allows more of the available DL constructors than the other two OWL profiles, but their usage is often restricted to one side of the inclusion axioms. This makes efficient RDF and rule reasoners applicable such as RDFox [64].

3   Nonmonotonic Rules

Nonmonotonic rules in Logic Programming (LP) have been intensively studied in the literature and a large body of theoretical results for different semantics of such rules has been presented (see e.g. [59]). Among them, the most widely used today are the answer set semantics [32] and the well-founded semantics [31], for which efficient implementations exist. In this section, we give a brief overview on nonmonotonic rules and these two semantics in particular, as they are essential for the integration of ontologies and rules.

In a broad sense, rules are used to represent that if a certain condition is true, then so is the indicated consequence. This includes the usage of default negation (in the condition) using operator **not**, which can be used to represent that something holds unless some condition is verified. We can for example state a rule that indicates that, for any vertebrate, the heart is on the left-hand-side of the body unless it is explicitly known to be on the right-hand-side:

$$HeartLeft(x) \leftarrow Vertebrate(x), \textbf{not } HeartRight(x) \qquad (7)$$

This rule is in principle applicable to arbitrary individuals due to the usage of variables, in this case x, and it intuitively allows us to infer for any x that $HeartLeft(x)$ holds unless $HeartRight(x)$ holds.

In terms of syntax, we note that the basic building blocks of rules are rather easy to define. All we need is a set of predicates of some arity, i.e., the number of arguments they admit, and a set of terms over constants and variables as these arguments. These, in turn, allow us to define rules as implications over such atoms possibly with default negation in the condition.

More formally, we consider disjoint sets of constants Σ_c, variables Σ_v, and predicate symbols Σ_p.[4] Then a *term* is a constant $c \in \Sigma_c$ or a variable $v \in \Sigma_v$, and an *atom* is of the form $p(t_1, \ldots, t_n)$, where p is an n-nary predicate symbol in Σ_p and the t_i, for $i = 1, \ldots, n$ and $n \geq 0$, are terms. Atoms with $n = 0$ are also commonly called propositional. Such atoms allow us to express that certain n-nary relations hold, or, using default negation **not**, that they do not hold in absence of information to the contrary. Atoms and default negated atoms are also called *literals*.

Logic programs then consist of rules that combine literals in logic formulas of a specific form. More precisely, a *disjunctive (logic) program* P consists of finitely many (implicitly universally quantified) *disjunctive rules* of the form

$$H_1 \vee \cdots \vee H_l \leftarrow A_1, \ldots, A_n, \textbf{not } B_1, \ldots, \textbf{not } B_m \qquad (8)$$

where H_k, A_i, and B_j are atoms. Such a rule can be divided into the *head* $H_1 \vee \cdots \vee H_l$ and the *body* $A_1, \ldots, A_n, \textbf{not } B_1, \cdots, \textbf{not } B_m$, where "," represents conjunction. Hence such a rule is to be read as "If A_1 and ... and A_n are true and B_1 and ... and B_m are not, then (at least) one of H_1 to H_l is".

We also identify the elements in the head and in the body of a rule with the sets $\mathcal{H} = \{H_1, \ldots, H_l\}$, $\mathcal{B}^+ = \{A_1, \ldots, A_n\}$, and $\mathcal{B}^- = \{B_1, \ldots, B_m\}$, where $\mathcal{B} = \mathcal{B}^+ \cup \textbf{not } \mathcal{B}^-$, and we occasionally abbreviate rules with $\mathcal{H} \leftarrow \mathcal{B}^+, \textbf{not } \mathcal{B}^-$ or even $\mathcal{H} \leftarrow \mathcal{B}$. Note that, in line with this set notation, the disjunction in the head and the conjunction in the body are commutative, i.e., the order of the elements in the head and in the body does not matter.

There are a number of different kinds of rules which yield a different expressiveness depending on how many and which literals are allowed in the head and the body of each rule, and we recall them in the following, as different integrations of ontologies and nonmonotonic rules admit different kinds of such rules.

---

[4] Please note that often also function symbols are introduced in the literature of LP, but since they jeopardize decidability of reasoning, and usually are not considered in integrations of ontologies and nonmonotonic rules, we do omit them here.

Normal rules do only admit one atom in the head, whereas *positive rules* do not admit default negation, and normal and positive programs can be defined accordingly. In particular, if the body of a rule is empty, i.e., $n = m = 0$, the rule is called a *fact*, and if, alternatively, the head is empty, i.e., $l = 0$, a *constraint*.

Constraints are an interesting modelling feature of nonmonotonic rules, as they allow to impose restrictions on the presence of certain information:

$$SSN\_OK(x) \leftarrow hasSSN(x, y) \tag{9}$$

$$\leftarrow Person(x), \mathbf{not}\ SSN\_OK(x) \tag{10}$$

This states that if x has a (known) social security number y, then x's social security number status is fine (9), and that there can be no person x whose social security number status is not fine (10). Note that the DL axiom (2) is not an alternative as it does not impose a restriction, but rather allows us to infer that there is a social security number, though we do not know which.

Here, unlike for DLs, no restriction on the syntax of predicates is made, which means that reasoning with such programs would be undecidable when working with an infinite domain. Commonly, this is prevented by ensuring that variables in the rules can only be instantiated with "known values", i.e., that are known in the program. Such rules are called *safe* which, formally, is the case if each variable in a rule of the form (8) occurs in an atom A_i for some i, $1 \leq i \leq n$, and we assume in the following that all rules are safe. For example, the rules (7), (9), and (10) are safe, whereas the rule

$$\leftarrow Person(x), \mathbf{not}\ hasSSN(x, y)$$

is not due to y not occurring in any positive literal in the rule body.

This restriction to known individuals seems arguably severe in comparison to DLs where we can reason over an infinite domain and unknown individuals. On the other hand, the following example rule taken from the context of Example 1

$$CompliantShpmt(x) \leftarrow ShpmtCommod(x, y), HTSCode(y, z),$$
$$ShpmtDeclHTSCode(x, z)$$

states that x is a compliant shipment, if x contains y whose harmonized tariff code is z and x is declared to contain z. This can be easily expressed as a rule, whereas a representation via DLs proves difficult due to the fact that, if viewed as a graph, the variable connections established in the rule provide one link from x to z via y and one direct link, which is not tree-shaped. Hence, both formalisms indeed differ as to what can be represented.

We next proceed by giving an overview on the two standard semantics for such nonmonotonic rules, answer set semantics and well-founded semantics.

3.1   Answer Set Semantics

Answer Set Programming (ASP) is a declarative programming paradigm tailored towards the solution of large combinatorical search problems. The central idea is to encode the given problem in a declarative way using rules and use one of the efficient answer set solvers available, such as clasp [29] or DLV [3], to determine the answer sets corresponding to the solutions of the problem. The approach builds on the answer set semantics [32] which is a two-valued nonmonotonic declarative semantics for logic programs with close connections to other logic-based formalisms such as SAT, Default Logic, and Autoepistemic Logic.

While the full ASP language comes with a number of additional syntactic constructs beyond the syntax of programs we have presented so far (cf. the ASP-Core-2 Input Language Format [18]), we limit our considerations to normal programs for the sake of readability and since this suffices to convey the main ideas.

The models considered in this semantics, called answer sets, are represented by a set of atoms occurring in a given program, that are true. Those atoms from the program not occurring in the answer set are false. Since rules may contain variables, rules are grounded first, i.e., all variables are instantiated with constants occurring in the program in all possible ways, and the set of all ground instances of the rules of a program P is denoted by $\mathsf{ground}(P)$.

Example 2. Consider program P consisting of rule $hasSSN(Bob, 123) \leftarrow$ together with (9). Then $\mathsf{ground}(P)$ consists of:

$$SSN\_OK(Bob) \leftarrow hasSSN(Bob, 123)$$
$$SSN\_OK(Bob) \leftarrow hasSSN(Bob, Bob)$$
$$SSN\_OK(123) \leftarrow hasSSN(123, Bob)$$
$$SSN\_OK(123) \leftarrow hasSSN(123, 123)$$
$$hasSSN(Bob, 123) \leftarrow$$

Of course, state-of-the-art ASP solvers will only keep the first of the grounded rules in such a situation, as the body atom in the other rules can never be true anyway. Either way, it is clear that $\{SSN\_OK(Bob), hasSSN(Bob, 123)\}$ is a model if we treat these rules as implications in first-order logic.

Now, the main idea of answer sets builds on guessing a model and checking that it satisfies a certain minimality criterion, namely, that there is some rule that supports the truth of some atom in an answer set (which is not the case, e.g., for $hasSSN(Bob, Bob)$ in Example 2). In more detail, based on the guessed model, a reduct of the (ground) program is created which does not contain default negation, and for which it can be checked whether the originally guessed model is a minimal model of the resulting reduct.

Given a program P and a set I of atoms, the *reduct* P^I [32] is defined as

$$P^I = \{\mathcal{H} \leftarrow \mathcal{B}^+ : \mathcal{H} \leftarrow \mathcal{B}^+, \mathbf{not}\ \mathcal{B}^- \text{in } \mathsf{ground}(P) \text{ such that } \mathcal{B}^- \cap I = \emptyset\}.$$

Intuitively, rules that contain an atom in I (assumed to be true) in the negative body are removed. In the remaining rules, all negated atoms are removed.

The resulting program is positive, and for normal programs, we can determine all necessary consequences of this program using the following operator T_P:

$$T_P(I) = \{H \mid H \leftarrow \mathcal{B} \in P \text{ and } \mathcal{B} \subseteq I\}$$

Basically, all rule heads are collected whose body atoms are true given I.

This operator can be iterated as follows, starting with $T_P \uparrow 0$, thus allowing to compute the deductive closure Cn of such a positive normal program P:

$$T_P \uparrow 0 = \emptyset \qquad T_P \uparrow (n+1) = T_P(T_P \uparrow n) \qquad Cn(P) = T_P \uparrow \omega = \bigcup_n T_P \uparrow n$$

Then, a set of atoms X is an *answer set* of program P iff $Cn(P^X) = X$.

For normal programs without negation, this closure in fact amounts to computing all necessary consequences of the given program. For example, for the ground program in Example 2 which does not contain default negation, clearly $\{SSN\_OK(Bob), hasSSN(Bob, 123)\}$ is this closure, hence its only answer set.

Example 3. Consider a normal program P containing just two rules.

$$HeartLeft(Bob) \leftarrow \textbf{not } HeartRight(Bob)$$
$$HeartRight(Bob) \leftarrow \textbf{not } HeartLeft(Bob)$$

If we consider $M_1 = \{HeartLeft(Bob), HeartRight(Bob)\}$, then $P^{M_1} = \{\}$, whose closure is $\emptyset$, hence M_1 is not an answer set. If we consider $M_2 = \{\}$, then $P^{M_2} = \{HeartLeft(Bob) \leftarrow, HeartRight(Bob) \leftarrow\}$ and the closure is $\{HeartLeft(Bob), HeartRight(Bob)\}$, so M_2 is not an answer set either. Now, consider $M_3 = \{HeartLeft(Bob)\}$. Then $P^{M_3} = \{HeartLeft(Bob) \leftarrow\}$, and M_3 is an answer set. The same is true for $M_4 = \{HeartRight(Bob)\}$ by symmetry.

Indeed, in general, a program may have several answer sets, which corresponds to the idea that combinatorial problems may have several solutions. In particular, a program may also have no answer sets – take $Person(Bob)$ together with rules (9) and (10). This is intended as we do not know the social security number of Bob, and in more general terms combinatorial problems may have no solution.

3.2   Well-Founded Semantics

The well-founded semantics [31] was developed around the same time as the answer set semantics, but focuses more on query answering over knowledge expressed with nonmonotonic rules. A central idea is to avoid computing entire models for a given program, but rather compute inferences in a top-down fashion, i.e., start with a query and compute only the part of a model necessary to obtain the answer. This is aligned with the ideas of Prolog [76], however unlike Prolog, the well-founded semantics is fully declarative and its major efficient implementation XSB Prolog [77] avoids undecidability caused by infinite loops while querying, and uses tabling to avoid unnecessary re-computation of already queried (intermediate) results.

Unlike the answer set semantics, the well-founded semantics is based on three-valued interpretations for which the set of truth values is extended by introducing a third truth value representing undefined. The basic idea behind this is to stay agnostic in situations where the knowledge encoded in a program permits a choice, such as between $HeartLeft(Bob)$ and $HeartRight(Bob)$ in Example 3, and rather leave both undefined. The benefit is that only a single well-founded model exists which can be efficiently computed in an iterative manner and for which corresponding querying procedures can be defined.

Formally, given a program P, a *three-valued interpretation* is a pair of sets of atoms (T, F) with $T \cap F = \emptyset$, where elements in T are mapped to true, elements in F are mapped to false, and the remaining to undefined. Such interpretations can also be represented as $T \cup \mathbf{not}\ F$. For example, using the latter notation, the well-founded model of the program in Example 3 can be represented as $\{\}$ corresponding to both $HeartLeft(Bob)$ and $HeartRight(Bob)$ being undefined, whereas the sets $\{HeartLeft(Bob), \mathbf{not}\ HeartRight(Bob)\}$ and $\{\mathbf{not}\ HeartLeft(Bob), HeartRight(Bob)\}$ correspond to the two answer sets in Example 3, respectively.

We can determine the truth value of a set of atoms involving undefined atoms by defining that, for an atom A undefined w.r.t. some three-valued interpretation, $\mathbf{not}\ A$ is again undefined and that for a conjunction of atoms, its truth value is the minimum of the truth values of the involved elements with respect to the order false < undefined < true.

Based on this, we can define the well-founded model and show how it can be computed. Here, we only do the latter and refer for the formal definition of the well-founded model to [31]. Essentially, the well-founded model can be computed by starting with an empty set (recall that this corresponds to everything is undefined), and iteratively add, based on the program, atoms that are necessarily true and necessarily false.

Regarding necessarily true information, the operator T_P defined previously for computing the closure Cn can be used, only now I is a three-valued interpretation.

For negative information, so-called unfounded sets are used, that refer to a set of atoms that given a program P and an interpretation I can never become true (nor undefined). Formally, such a set U is unfounded w.r.t. program P and interpretation I if each atom $A \in U$ satisfies the following condition. For each rule $A \leftarrow B$ in $\mathsf{ground}(P)$ at least one of the following holds:

(Ui) Some literal in B is false in I.
(Uii) Some (non-negated) atom in B occurs in U.

The idea is that A is unfounded if all rules with head A either contain some false literal in the body or an atom which is also unfounded. For example, given $I = \emptyset$ and program P composed of the two rules:

$$hasSSN(Bob, 123) \leftarrow isSSNOf(123, Bob)$$
$$isSSNOf(123, Bob) \leftarrow hasSSN(Bob, 123)$$

$\{hasSSN(Bob, 123), isSSNOf(123, Bob)\}$ is an unfounded set w.r.t. P and I.

The union of all unfounded sets of P w.r.t. I is the greatest unfounded set, denoted $U_P(I)$ which together with T_P can be used to define an operator W_P for programs P and interpretations I:

$$W_P(I) = T_P(I) \cup \textbf{not}\ U_P(I).$$

This operator can be iterated to obtain the well-founded model M_{wf}, i.e., all atoms that are necessarily true and false, respectively.

$$W_P \uparrow 0 = \emptyset \qquad W_P \uparrow (n+1) = W_P(W_P \uparrow n) \qquad M_{wf}(P) = W_P \uparrow \omega = \bigcup_n W_P \uparrow n$$

Example 4. Consider the following program P.

$$Person(Bob) \leftarrow$$
$$hasSSN(Bob, 123) \leftarrow$$
$$SSN\_OK(x) \leftarrow hasSSN(x, y)$$
$$check(x) \leftarrow Person(x), \textbf{not}\ SSN\_OK(x)$$

We obtain:

$$W_P \uparrow 0 = \emptyset$$
$$W_P \uparrow 1 = \{Person(Bob), hasSSN(Bob, 123)\}$$
$$W_P \uparrow 2 = W_P \uparrow 1 \cup \{SSN\_OK(Bob)\}$$
$$W_P \uparrow 3 = W_P \uparrow 2 \cup \{\textbf{not}\ check(Bob)\}$$

An alternative equivalent definition exists, called the alternating fixed-point [30], which computes this model based on the reduct of the program used for determining answer sets, and SLG resolution [20] provides a corresponding top-down procedure for the well-founded semantics implemented in XSB.

Regarding a comparison between the well-founded semantics and the answer set semantics, we note that in terms of computational complexity, the former is preferrable as the unique well-founded model can be computed in polynomial time (w.r.t. data complexity, i.e., only the size of the number of facts varies), whereas guessing and checking answer sets is at least in NP. In addition, when querying, we only require the part of the knowledge base that is relevant for the query, which aids efficiency in comparison to answer sets. For example, when querying for the medication of a specific patient, we certainly do not care about the medication of possibly thousands of other patients. Moreover, the well-founded model (for normal programs) always exists. On the other hand, if the considered problem is at least partially combinatorial, answer sets are clearly preferred. In fact, the answer set semantics is more expressive in general: take the two rules from Example 3 together with the following two rules.

$$LivingBeing(Bob) \leftarrow HeartLeftBob)$$
$$LivingBeing(Bob) \leftarrow HeartRight(Bob)$$

Then, in the unique well-founded model everything is undefined, whereas both answer sets contain *LivingBeing(Bob)*. Thus, ultimately the choice between the depends on the intended application.

4 How to Integrate Ontologies and Rules?

Having reviewed the two formalisms, DL ontologies and nonmonotonic rules in detail, and their differing characteristics and benefits w.r.t. what kind of knowledge can be represented and reasoned with, the question arises what to do if we want to use the favorable characteristics of both simultaneously.

This question has been tackled in the literature and a plethora of different approaches has been presented. Discussing all these proposals here in detail would not be possible for the sheer amount of them, however, common characteristics and criteria have emerged along which these proposals have been defined, and we want to discuss these here to provide a better idea on the main considerations to take into account when providing such an integration.

Before we delve into this, let us look at a concrete, larger example that illustrates the benefits of such an integration with more detail.

Example 5. Recall the setting described in Example 1 on a customs service needing to assess incoming cargo for risks based on a variety of information. Figure 3 shows a DL ontology $\mathcal{O}$ and a set of nonmonotonic rules $\mathcal{P}$ containing part of such information that we explain in more detail.

The ontology $\mathcal{O}$ contains a classification of commodities based on their harmonized tariff information (HTS chapters, headings and codes)[5], a taxonomy of commodities (here exemplified using special kinds of tomatos), together with indications on tariff charges depending on the kind of product and their packaging, as well as a geographic classification, along with information about producers who are located in various countries.

The program $\mathcal{P}$ contains data on the shipments. Here, a shipment has several attributes: the country of origin, the commodity it contains, its importer and producer, exemplified by (partial) information about three shipments, s_1, s_2 and s_3. For importers, it also indicates if they have a history of transgressing the regulations.

Then there is a set of rules for determining what to inspect. The first rules provide information about importers, namely whether they are admissible (if they are not known to be registered transgressors), for which kind of products they are approved, and whether they are expeditable (combining the former two). The rules also allow us to associate a commodity with its country of origin (where it shipped from), and determining whether a shipment is compliant, i.e., if there is a match between the filed cargo codes and the actually carried commodities.

The final three rules serve the overall task to access all the information and assess whether some shipment should be inspected in full detail, partially, or not at all. In particular, at least a partial inspection is required whenever

---

[5] This is adapted from https://hts.usitc.gov/.

* * * $\mathcal{O}$ * * *

Commodity ≡ (∃HTSCode.⊤)   EdibleVegetable ≡ ∃HTSChapter.{'07'}
CherryTomato ≡ ∃HTSCode.{'07022'}   Tomato ≡ ∃HTSHeading.{'0702'}
GrapeTomato ≡ ∃HTSCode.{'07021'}   Tomato ⊑ EdibleVegetable
CherryTomato ⊑ Tomato   GrapeTomato ⊑ Tomato
CherryTomato ⊓ Bulk ≡ ∃TariffCharge.{'$0'} CherryTomato ⊓ GrapeTomato ⊑ ⊥
GrapeTomato ⊓ Bulk ≡ ∃TariffCharge.{'$40'} Bulk ⊓ Prepackaged ⊑ ⊥
CherryTomato ⊓ Prepackaged ≡ ∃TariffCharge.{'$50'}
GrapeTomato ⊓ Prepackaged ≡ ∃TariffCharge.{'$100'}
EURegisteredProducer ≡ ∃RegisteredProducer.EUCountry
LowRiskEUCommodity ≡ (∃ExpeditableImporter.⊤) ⊓ (∃CommodCountry.EUCountry)

EUCountry($portugal$)   EUCountry($slovakia$)

* * * $\mathcal{P}$ * * *

ShpmtCommod(s_1, c_1)   ShpmtDeclHTSCode(s_1, h7022)
ShpmtImporter(s_1, i_1)   CherryTomato(c_1)   Bulk(c_1)
ShpmtCommod(s_2, c_2)   ShpmtDeclHTSCode(s_2, h7021)
ShpmtImporter(s_2, i_2)   GrapeTomato(c_2)   Prepackaged(c_2)
ShpmtCountry($s_2, slovakia$)   RegisteredTransgressor(i_1).
ShpmtCommod(s_3, c_3)   ShpmtDeclHTSCode(s_3, h7022)
ShpmtImporter(s_3, i_3)   GrapeTomato(c_3)   Bulk(c_3)
ShpmtCountry($s_3, portugal$)   ShpmtProducer(s_3, p_1)
ShpmtCountry($s_1, portugal$)   RegisteredProducer($p_2, slovakia$)

AdmissibleImporter($\mathbf{x}$) ← ShpmtImporter($\mathbf{y, x}$), **not** RegisteredTransgressor($\mathbf{x}$)
ApprovedImporterOf($i_2, \mathbf{x}$) ← EdibleVegetable($\mathbf{x}$)
ApprovedImporterOf($i_3, \mathbf{x}$) ← GrapeTomato($\mathbf{x}$)
ExpeditableImporter($\mathbf{x, y}$) ← ShpmtCommod($\mathbf{z, x}$), ShpmtImporter($\mathbf{z, y}$),
                        AdmissibleImporter($\mathbf{y}$), ApprovedImporterOf($\mathbf{y, x}$)
CommodCountry($\mathbf{x, y}$) ← ShpmtCommod($\mathbf{z, x}$), ShpmtCountry($\mathbf{z, y}$)
CompliantShpmt($\mathbf{x}$) ← ShpmtCommod($\mathbf{x, y}$), HTSCode($\mathbf{y, z}$), ShpmtDeclHTSCode($\mathbf{x, z}$)
PartialInspection($\mathbf{x}$) ← ShpmtCommod($\mathbf{x, y}$), **not** LowRiskEUCommodity($\mathbf{y}$)
FullInspection($\mathbf{x}$) ← ShpmtCommod($\mathbf{x, y}$), **not** CompliantShpmt($\mathbf{x}$)
FullInspection($\mathbf{x}$) ← ShpmtCommod($\mathbf{x, y}$), Tomato($\mathbf{y}$), ShpmtCountry($\mathbf{x}, slovakia$)

Fig. 3. Ontology $\mathcal{O}$ and nonmonotonic rules $\mathcal{P}$ for cargo assessment.

the commodity is not a LowRiskEUCommodity based on inferences in the DL part, which n turn requires assessing further information in the rules, namely CommodCountry and ExpeditableImporter. A full inspection is required if a shipment is not compliant or if some suspicious cargo is observed, in this case tomatoes from slovakia (you may imagine for the sake of the example that we are in the winter period).

Note that the example indeed utilizes the features of rules and ontologies: for example, exceptions to the partial inspections can be expressed, but at the same time, taxonomic and non-closed knowledge is used, e.g., some shipment may in fact originate from the EU, but this information is just not available.

To be able to obtain the desired conclusions from the combination of the knowledge bases written in these two formalisms, we need to find ways to integrate them. As already mentioned, many proposals exist in the literature, defined along the lines of guiding principles which we now want to discuss. In the course of this discussion, often, these principles are inspired in their formulation by [62], where they were used for arguing in favor of a specific approach. Here, in particular when there are several options w.r.t. a certain characteristic, it is our stance not to argue in favor of a certain solution, but rather discuss their corresponding advantages.

Faithfulness. The first desirable property we want to discuss is Faithfulness, i.e., the idea that the integration of DLs and nonmonotonic rules should preserve the semantics of both its base formalisms. In other words, the addition of rules to a DL should not change the semantics of the DL and vice-versa. In particular, when either of the two components is empty, the semantics should simply be the one of the base formalism of the non-empty component. This is beneficial for two reasons. First, it eases its adoption for knowledge engineers knowledgable in (at least) one of the base formalisms, that want to augment an ontology with rules or vice-versa. Second, and maybe even more important, this is crucial to facilitate re-using existing state of the art reasoners for each of the components, reducing considerably the necessary effort to provide implementations. Thus, without going into details of an actual faithful integration, for reasoning with the knowledge bases presented in Example 5, it would in principle suffice to choose a DL reasoner appropriate for the expressiveness used in the DL and an ASP solver (or XSB depending on the desired rule semantics) and determine an interface between the two to obtain all desired inferences.

Tightness. The characteristic of tightness relates to the structure/topology of the integration in the sense that whether conclusions resulting from one of the formalisms can be used to derive further conclusions in the other. Clearly, there are several possible solutions. One option is to layer one formalism on top of the other, in the sense that conclusions of one approach can be used by the other, but not vice-versa. Such a solution is certainly easier to handle on the technical level, since, assuming a faithful integration, this way, we can simply first compute the conclusions within the formalism on the lower level, and pass these to the formalism on the upper level to compute the conclusions there. On the other hand, a tight integration requires that neither of the two formalisms is layered on top of the other, rather, both the DL and the rule component should be able to contribute to the consequences of the other component. This is technically more advanced, but has the advantage that it can easily cover examples such as the one presented in Example 5, where, e.g., LowRiskEUCommodity is used in the rules, but derived in the ontology, which in turn requires inferences from the rules. Of course, in applications where such tight integration is not necessary, a layered solution suffices. On the other hand, a tight integration is not subject to problems when later introducing transfer of inferences from one component to the other that was previously not present. Finally, there are also intermediate solutions in which inferences can be passed between both components of the integration, but under some certain restrictions.

DL Reasoning View. This characteristic refers to what kind of inferences are passed from the DL component to the rule component of the integration. Clearly, for layered approaches in which rules are on the lower layer, this characteristic is irrelevant. However, the vast majority of approaches actually do pass conclusions from the DL component to the rule component. The essential question is, given a DL ontology, do we consider inferences on the level of individual models or on the level of consequences, i.e., truth in all models? In other words, are the inferences we pass from the ontology to the rules model-based or consequence-based, i.e., is the integration a world-centric or entailment-centric approach (as alternatively termed in [56])? The benefit of model-based inferences is that they allow for a higher expressiveness in the sense that possible alternatives present in the ontology can be passed to the rules resulting in more inferences there as well. On the downside, such alternatives based on different models may not have strong support as a conclusion obtained from the ontology. For consequence-based approaches, the situation is exactly the converse. Any inference is true in all models, but this limits the extent to which inferred information is passed from the ontology to the rules. This is ultimately actually similar to the idea of brave vs. cautious reasoning in logics, i.e., truth in one or all models, and which one of the two views is adapted depends on the concrete application at hand. For Example 5, this depends on whether we want to impose inspections based on something possibly being true or an entailment, i.e., being true in all models. For example, LowRiskEUCommodity should probably only be considered under consequences, as discarding a partial inspection based on the fact that the commodity in question may represent a lower risk, is probably not a good idea.

Flexibility. This characteristic deals with the question whether the same predicate can be viewed under both the open and closed world assumption. The central idea is that a flexible approach allows us to enrich a DL ontology with non-monotonic consequences from the rules, and, conversely, to enrich the rules with the capabilities of ontology reasoning described by a DL ontology. This allows us to distinguish between approaches that fix to which predicates either OWA or CWA is applied and those that do not. Again, a flexible approach is commonly more expressive, possibly at the expense of requiring a technically more advanced integration, since separate languages for the two components facilitate the re-use of existing semantics for the individual components. Yet, this may require some additional effort to ensure that corresponding predicates (between the ontology and the rules) are appropriately synchronized. In the case of Example 5, an approach capturing it as is needs to be flexible as the predicates appear simultaneously in both components.

Decidability and Complexity. Finally, a formalism that can be used in applications should be at least decidable, and preferably of low worst-case complexity. Given the huge amount of data (on the Web), it is clearly preferable to have a system that is not only decidable but also computationally as efficient as possible. Still, this needs to take into account the tradeoff between expressiveness of the formalism and complexity of reasoning, and in certain situations, the required expressiveness is more important. In any case, both base formalisms come with established methods to ensure decidability, and this should preferably be maintained in their integration. The exact computational complexity for desired reasoning tasks then depends on the expressiveness and semantics applied in the two base formalisms and which of the beforementioned characteristics have been adopted.

5   Concrete Integrations

Having reviewed the main guidelines/criteria along which the many existing approaches for integrating ontologies and rules have been developed, in this section, we want to give an overview of some of the more prominent such proposals. While the question of prominence to some extent certainly is in the eye of the beholder, and though we cannot discuss all existing approaches here in more detail, we believe that the provided selection covers different aspects w.r.t. the previously mentioned criteria and more details on the remaining approaches can be obtained from pointers to the literature.

5.1 $\mathcal{DL}$+log

One of the first combinations of non-monotonic rules and ontologies is called r-hybrid knowledge bases [67], which subseqently has been extended to $\mathcal{DL}$+log [68]. $\mathcal{DL}$+log combines disjunctive Datalog (consisting of rules of the form (8)) with an arbitrary (decidable) DL.

An essential idea is to separate the admitted predicates into DL predicates and non-DL predicates, i.e., those that can appear in the DL part (and in the rules), and those that only appear in the rules. A knowledge base $\mathcal{K}$ in this formalism consists of an ontology $\mathcal{O}$ and a set of rules $\mathcal{P}$, where each rule r is of the following form:

$$p_1(\boldsymbol{x}_1) \vee \ldots \vee p_n(\boldsymbol{x}_n) \leftarrow r_1(\boldsymbol{y}_1), \ldots, r_m(\boldsymbol{y}_m), s_1(\boldsymbol{z}_1), \ldots, s_k(\boldsymbol{z}_k),$$
$$\mathbf{not} \; u_1(\boldsymbol{w}_1), \ldots, \mathbf{not} \; u_h(\boldsymbol{w}_h)$$

such that $n \geq 0$, $m \geq 0$, $k \geq 0$, and $h \geq 0$, the $\boldsymbol{x}_i$, $\boldsymbol{y}_i$, $\boldsymbol{z}_i$, and $\boldsymbol{w}_i$ are tuples of variables and constants, each p_i a DL or a non-DL predicate, each s_i a DL predicate, and each r_i and u_i a non-DL predicate. Additionally, each variable occurring in a rule must appear in some y_i or z_i (rule safety – similar to what we have seen in Sect. 3) and every variable appearing in some x_i of r must appear in at least one of the y_i (weak safety). The latter notion is weaker/less restrictive than DL-safety [63] applied for r-hybrid knowledge bases, which requires that every variable (and not just those in x_i) in the rule occurs in at least one y_i, since there may exist variables that only appear in a DL-atom in the body of a rule. For example, we may assume that, in Example 5, data on registered producers is also stored in the ontology (in the ABox), i.e., EURegisteredProducer is a DL-predicate, and adapt the DL axiom on EURegisteredProducer as a rule:

$$\text{EURegisteredProducer}(\mathbf{x}) \leftarrow \text{RegisteredProducer}(\mathbf{x}, \mathbf{y}), \text{EUCountry}(\mathbf{y}) \qquad (11)$$

Then, this rule is weakly-safe, but not DL-safe, as, according to our assumptions, $\mathbf{x}$ does not occur in the rule body in an atom built over a non-DL predicate (nor does $\mathbf{y}$). This is interesting as it allows one to pose arbitrary conjunctive queries over DL predidates to the DL component, for which there are known algorithms for many DLs [57]. The standard rule safety nevertheless ensures that there is no variable only appearing in a default negated atom.

The main idea of the semantics for $\mathcal{DL}+\log$ [68] is based on the definition of a projection, similar in spirit to the idea of the reduct for answer sets, which intuitively simplifies the program by evaluating the DL-atoms. Basically, given an interpretation I, rules with DL-atoms that conflict with I are deleted and the remaining DL-atoms are omitted. The resulting program is free of DL-atoms, and $\mathcal{I}$ is a model of $\mathcal{K}$ if $\mathcal{I}$ is a model of $\mathcal{O}$, and $\mathcal{I}$ restricted to the non-DL predicates is an answer set of the projected program. This way, DL atoms are evaluated under the open world assumption in the spirit of conjunctive queries to the DL part, whereas non-DL atoms are evaluated under the closed world assumption.

Example 6. Consider a simple example composed of rule (11) together with two ABox assertions RegisteredProducer(s_4, GB), CountryIE(GB) and an axiom

$$\text{CountryIE} \sqsubseteq \text{EUCountry} \sqcup \text{NonEUCountry}$$

representing that any country in Europe is either in the European Union or not, and that s_4 is a registered producer in Great Britain, which is a country in Europe. Then there are two models:

$$M_1 = \{\text{RegisteredProducer}(s_4, GB), \text{CountryIE}(GB), \text{EUCountry}(GB),$$
$$\text{EURegisteredProducer}(s_4)\}$$
$$M_2 = \{\text{RegisteredProducer}(s_4, GB), \text{CountryIE}(GB), \text{NonEUCountry}(GB)\}$$

Notably, M_1 and M_2 correspond to the essential two models of $\mathcal{O}$, and, for M_2, the evaluation of DL-atoms removes all instances of (11).

This then allows the definition for reasoning algorithms along the idea of guessing and checking interpretations based on these projections, and it has been shown that, in combination with $\mathcal{O}$ in DL-Lite, the DL underlying OWL 2

QL, the data complexity does not increase with respect to the data complexity of the rule component alone.

The semantics is faithful, tight, and decidable (due to the safety restrictions), but it is not flexible, as non-monotonic reasoning is by definition restricted to non-DL atoms, i.e., no default reasoning over information appearing in the DL is possible. Finally, we note that this approach is indeed model-based, since it suffices that a considered interpretation $\mathcal{I}$ models $\mathcal{O}$, without considering entailment, i.e., truth in all models of $\mathcal{O}$.

5.2   Dl-Programs

Another important approach is called description logic programs (dl-programs) [27]. Here, a knowledge base $\mathcal{K}$ consists of a DL ontology $\mathcal{O}$ and a program $\mathcal{P}$ containing (non-disjunctive) dl-rules of the form

$$H \leftarrow A_1, \leftarrow A_n, \text{not } B_1, \ldots, \text{not } B_m$$

where H is a first-order atom,[6] and all A_i and B_j are first-order atoms or special dl-atoms to query the ontology. Similarly to $\mathcal{DL}+\log$, the set of predicates is divided into disjoint sets of DL predicates, which only appear in dl-atoms and in $\mathcal{O}$, and non-DL predicates, that only appear in the rules. Such dl-atoms are meant to serve as interfaces between the ontology and the rules, allowing that information derivable in the DL can be queried for in the rules, while information in the rules can be passed to the DL in the course of this process.

As a simple example, consider that the rules contain a fact p(Bob), representing that Bob is a person, together with ontology axioms (5) and (6), from which we can derive that every person is a vertebrate. Note that the ontology and the rules use a different predicate to represent the concept of person. Then, a dl-atom $DL[\text{Person} \uplus \text{p}; \text{Vertebrate}](\text{Bob})$ represents a query for Vertebrate(Bob) where Person in the ontology is enriched with the inferences for p from the rules.

More formally, such dl-atoms are of the form

$$DL[S_1 \text{ } op_1 \text{ } p_1, \ldots, S_l \text{ } op_l \text{ } p_l; Q](t)$$

where S_i are DL predicates, p_i are non-DL predicates, $op_i \in \{\uplus, \cup, \cap\}$, Q is an n-ary DL predicate, and t a vector of n terms. Such dl-atoms are used to query the ontology for $Q(t)$ where certain ontology predicates S_i are altered by information derived in the rules. Intuitively, $\uplus$ is used to augment S_i with the derived knowledge from p_i; $\cup$ augments $\neg S_i$ with the derived knowledge from p_i and $\cap$ augments $\neg S$ with what is not derived in p_i. As queries, concept inclusions, negations of concept inclusions, concept and role assertions, and equalities and inequalities are allowed.

It should be noted that the transfer from the rules to the ontology is limited though, in the sense that it is not stored. This means that if a certain piece of information from the rules is to be used for each dl-query, then it has to be added explicitly in every dl-atom.

---

[6] Classical negation is also allowed, but we simplify here for the sake of presentation.

Example 7. The rule from Example 5 for PartialInspection can be adapted as:

PartialInspection($\mathbf{x}$) ← ShpmtCommod($\mathbf{x}, \mathbf{y}$),

not DL[ExpeditableImporter ⊎ e, CommodCountry ⊎ c; LowRiskEUCommodity]($\mathbf{y}$)

where e and c are non-DL predicates which replace ExpeditableImporter and CommodCountry in all the rules. In fact, to make this fully correct with the idea of dl-programs, all predicates that appear, in Example 5, in the ontology and in the rules need to be duplicated. This may look cumbersome at first glance, but it allows one to use the two components in a modular fashion which facilitates implementations, and reduces transfer of knowledge to what is necessary aiding efficiency.

Two different answer set semantics are defined slightly varying on how dl-atoms are preprocessed in the reduct. Both semantics may however create non-minimal answer sets which is why it is recommended to either use canonical answer sets based on loop formulas [80] or not to use the operator ∩ at all.

In addition, a corresponding well-founded semantics is defined for dl-programs [26] omitting the operator ∩, which builds on an extension of the unfounded sets construction to such dl-programs.

The approach itself is faithful for both versions, the answer set semantics and the well-founded semantics. The interaction with the DL component is consequence-based as the evaluation of queries in the dl-atoms is realized based on entailments, i.e., truth in all models of $\mathcal{O}$. Due to the specific interfaces, the dl-atoms, this approach is tight and flexible to a limited extent: rules cannot derive new facts about DL predicates, they can only pose conditional queries to the DL, although the operators in dl-atoms provide means to transfer knowledge temporarily. Moreover, we can never derive nonmonotonic consequences for DL predicates directly. Rather, we have to use the interfaces, which may appear under default negation. It has also been shown that dl-programs are decidable provided the considered DL is decidable. For the answer set semantics, in general the computational complexity increases that of the individual components (depending on the DL used and the kind of rules permitted), whereas for the well-founded semantics this can commonly be avoided. In particular for polynomial DLs, such as the tractable OWL profiles, dl-programs also maintain a polynomial data complexity. While the modular solution limits the approach in terms of tightness and flexibility, it facilitates its implementation. A prototype for both semantics has been defined that utilizes RACER [37] for DL reasoning and DLV [3,53] for the rule processing (where the well-founded semantics is implemented based on the alternating fixpoint). This has been subsequently generalized in hex-programs [24,66] where interfaces in the line of dl-atoms can be created to arbitrary external sources, thus further widening the applicability.

5.3 Hybrid MKNF

Hybrid MKNF knowledge bases [62] build on the logic of minimal knowledge and negation as failure (MKNF) [54], which corresponds to first-order logic extended by two modal operators **K** and **not**, that allow us to express that something is known and not known, respectively. Hybrid MKNF KBs essentially consist of two components: a first-order theory, in particular a DL ontology $\mathcal{O}$ (translatable into first-order logic), and a finite set of rules (similar to rules in ASP) over so-called modal atoms of the form

$$\mathbf{K}H_1 \vee \ldots \vee \mathbf{K}H_l \leftarrow \mathbf{K}A_1, \ldots, \mathbf{K}A_n, \mathbf{not}\ B_1, \ldots, \mathbf{not}\ B_m$$

where the H_i, A_j, and B_k are first-order formulas. The essential idea is that such a rule is to be read as "If all A_j are known to hold (in the sense of truth in all models), and all B_k are not known to hold, then one of the H_i is known to hold" (i.e., one of the H_i is true in all models). As modal operators are not admitted in the ontology $\mathcal{O}$, reasoning can still be applied within $\mathcal{O}$ on a per model basis as intended. Admitting, in the rules, first-order formulas within the scope of modal operators raises the expressivity, and allows, in such modal atoms, conjunctive queries to the ontology.

In fact, even more general MKNF$^+$ knowledge bases are considered in which rules may contain atoms not in scope of any modal operator. These can however be transformed into ordinary modal atoms provided the considered DL language is expressive enough to encode these first-order atoms with an equivalence to a new predicate used then in the rules in each such case.

Atoms are again divided into DL atoms and non-DL atoms, and for decidability, it is required that DL-safety holds, i.e., all variables occur in at least one non-DL atom, and that reasoning in the DL language together with the generalized atoms is decidable (the latter condition simplifies if no first-order formulas occur in the rules). Note that, due to DL-safety, rules such as (11) cannot be used, but this can be amended using a conjunctive query in a modal atom:

KEURegisteredProducer(x) ← **K**[∃y.(RegisteredProducer(x, y), ∧EUCountry(y))]

The semantics of Hybrid MKNF knowledge bases is given by a translation into an MKNF formula, i.e., a formula over first-order logic extended with two modal operators **K** and **not**. This translation, essentially, conjoins all rules with the first-order translation of $\mathcal{O}$ within scope of a single modal operator **K**. The (integrating) semantics for such formulas is defined based on a model notion over sets of interpretations. Such models contain all interpretations that model a given formula, minimizing what necessarily must be known to hold, in the sense that a larger set of models contains less atoms that are true in all models, hence the name minimal knowledge (and negation as failure).

Example 8. Consider a simple example adapted from Example 5 containing just CherryTomato $\sqsubseteq$ Tomato and **K**Tomato(o_1) ←. The corresponding formula in MKNF is $\mathbf{K}(\forall x \text{CherryTomato}(x) \rightarrow \text{Tomato}(x)) \wedge \mathbf{K}\text{Tomato}(o_1)$. Then, $M_1 =$ {{Tomato(o_1)}, {Tomato(o_1), CherryTomato(o_1)}} contains sets that model the formula. So does $M_2 =$ {{Tomato(o_1), CherryTomato(o_1)}}, but M_2 is not the maximal such set of sets. In fact, in M_2, CherryTomato(o_1) is true in all models, i.e., according to this model, **K**CherryTomato(o_1) holds, which clearly should not be an inference of the given knowledge base.

Since such a model representation is cumbersome and in general infinite, for reasoning, a finite representation based on modal atoms is provided that indicates which modal atoms occurring in the program are true and false. This then allows one to use algorithms whose ideas are closely aligned with that of answer sets – guess the true modal atoms and check whether certain conditions are verified, namely that the result is a minimal model for the formula. For Example 8, this amounts to determining that $\mathbf{K}\mathsf{Tomato}(o_1)$ is true.

A well-founded semantics based on alternating fixpoints is defined in [48] for hybrid MKNF knowledge bases, where no disjunction is allowed in the rule heads and only ordinary atoms are permitted in rules, i.e., no first-order formulas.

Due to the seemless embedding in the underlying unifying formalism, the approach is naturally flexible and tight. This allows us for example to capture Example 5, by simply introducing the modal $\mathbf{K}$ operators in all the rules (the default **not** is already present). It also is faithful with the respective underlying base formalisms and decidable, provided the required restrictions are satisfied. In terms of complexity, the semantics based on answer sets does in general increase the data complexity of its constituents, only under considerable restrictions on the rules, this can be avoided. For the well-founded semantics on the other hand, and similar to dl-programs, the complexity reduces in comparison, and for polynomial DLs, polynomial data complexity can be ensured for the integrated formalism (for computing the model as well as answering safe queries). Moreover, the approach is consequence-based as DL atoms appear in scope of modal operators, hence they are verified to be known, i.e., true in all models (of $\mathcal{O}$).

The approach is indeed very general, and MKNF$^+$ knowledge bases allow us to cover many approaches in the literature [62], including the ones discussed so far, in the sense that $\mathcal{DL}$+log KBs and dl-programs without the operator $\cap$ can be encoded into an equisatisfiable hybrid MKNF knowledge base.

In the latter case, a complementary formal result [28] shows that DL-safe and ground hybrid MKNF KBs can be embedded into dl-programs essentially by establishing the necessary transfer of information from rules to ontologies in every single DL-atom. This confirms that Example 5 which can be straightforwardly handled in hybrid MKNF, is also capturable in the case of dl-programs. The general idea is to just introduce one new auxiliary predicate for each DL atom, basically creating a program and a DL version of each such predicate, and then use dl-atoms in the bodies of rules (instead of DL atoms), to query the ontology, where, in the dl-atoms, the inferred information for all such DL atoms from the rules is passed to the corresponding DL component each time.

5.4   Resilient Logic Programs

Resilient Logic Programs (RLPs) [56] have been recently introduced with the aim to overcome a limitation of the existing approaches of integrations of ontologies and rules, namely the fact that, for obtaining inferences from the ontology, they use either model-based reasoning (e.g., $\mathcal{DL}$+log) or consequence-based reasoning (e.g., dl-programs and Hybrid MKNF), but not both. There are however problems where this is not sufficient, and the authors argue that integration solutions should be resilient in various scenarios.

As an example, consider that we are given a set of nodes, and we want to determine a directed graph G such that removing one arbitrary node from G will always result in a strongly connected graph (i.e., every vertex is reachable from every other vertex). In this case, the choice of which node is removed can be modeled in the ontology under the model-based view (one model per possible removed node), whereas we can use the rules to verify whether the corresponding resulting graph is strongly connected. However, the reachability relation involved in this check varies for different chosen nodes, which requires a universal quantification over the choices in the ontology (aligned with the consequence-based view).

A resilient logic program then is a tuple $\Pi = (\mathcal{P}, \mathcal{O}, \Sigma_{out}, \Sigma_{owa}, \Sigma_{re})$ consisting of a program $\mathcal{P}$, an ontology/first-order theory $\mathcal{O}$, and a distinct partition of the predicates occurring in $\mathcal{P}$ and $\mathcal{O}$, into output predicates Σ_{out}, open predicates Σ_{owa}, and response predicates Σ_{re}, such that response predicates are not allowed in $\mathcal{O}$, and where output predicates and response predicates are closed.

The semantics is defined in a way that can be viewed as a negotiation between $\mathcal{P}$ and $\mathcal{O}$. An answer set I over Σ_{out} needs to be determined that a) can be extended into a model of $\mathcal{O}$ by interpreting the predicates in Σ_{owa}, and b) no matter how I is extended into a model of $\mathcal{O}$, there always is a corresponding interpretation of the response predicates, which together with I is justified by $\mathcal{P}$ (in the sense of minimality/support by a rule as argued before, e.g., for answer set programs). This semantics uses a reduct inspired by [9,67], which handles default negated atoms in the rules as usual, but in addition also treats atoms based on open predicates in a similar fashion.

Example 9. Consider the following formalization of the graph problem taken from [56], where $\mathcal{O}$ is represented as a first-order theory.

For nodes $n_1, \ldots, n_k$, let $\Pi = (\mathcal{P}, \mathcal{O}, \Sigma_{out} = \{V, E\}, \Sigma_{owa} = \{in, out\}, \Sigma_{re} = \{\bar{E}, R\})$, where

$$\mathcal{O} = \{\exists x.out(x) \qquad \forall x.(V(x) \to (in(x) \vee out(x))),$$
$$\forall x.(V(x) \to (\neg in(x) \vee \neg out(x))),$$
$$\forall x \forall y.((out(x) \wedge out(y)) \to x = y)\}$$
$$\mathcal{P} = \{V(n_1), \ldots V(n_k), \qquad E(x,y) \vee \bar{E}(x,y) \leftarrow V(x), V(y),$$
$$R(x,z) \leftarrow R(x,y), R(y,z),$$
$$R(x,y) \leftarrow E(x,y), \textbf{not } out(x), \textbf{not } out(y),$$
$$\leftarrow V(x), V(y), x \neq y, \textbf{not } out(x), \textbf{not } out(y), \textbf{not } R(x,y)\}$$

Essentially, $\mathcal{O}$ allows us to choose exactly on vertex that is removed (out), whereas $\mathcal{P}$ provides possible directed edges between the given vertices ($E(x,y)$ corresponding to a chosen edge for vertices x and y, $\bar{E}(x,y)$ representing one not chosen), determines reachability (using $R(x,y)$), and a constraint checking that any remaining vertex can reach any other remaining vertex. The semantics

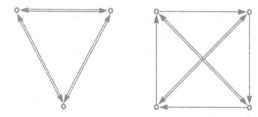

Fig. 4. A graph representation of the unique solution for Example 9 for 3 nodes, and one possible solution for 4 nodes.

of RLPs then only admits the desired solutions in which removing any single node results in a strongly connected graph. Figure 4 shows a representation of the only possible such graph with three nodes and one possible solution with four nodes. It is easy to see that removing any of the directed edges will result in a graph for which the required condition is no longer satisfied.

A more extensive example is presented in [56], where a company wants to process a fixed amount of customer orders per day. The exact configuration of the orders is not known in advance, and the objective is to determine those services to offer so that independently of the actual configurations of the orders, the tasks can be assigned to employees so that each task is completed by the end of the day. Here, the offered services are captured by the output predicates, possible configurations are modelled in the first-order theory, and the answer sets correspond to the viable schedules in which tasks are completed in time.

Decidability in RLPs is achieved by ensuring DL-safety [63], here w.r.t. output and response predicates, and requiring that for the (DL) theory, satisfiability under closed predicates must be decidable.

It is shown that disjunctive programs can be embedded into RLPs. Also, under some additional restrictions, namely limiting default negation for response predicates and restricting theories to correspond to positive disjunctive rules, RLPs can be translated into disjunctive ASP which allows for the usage of state-of-the-art ASP solvers when reasoning. A reduction to $\exists\forall\exists$-quantified Boolean formulas is given, which shows that the computational complexity is higher than that of answer sets, which is also confirmed when using concrete DLs of different expressiveness. In the context of these concrete DLs, a relaxed safeness condition is also provided that admits safety with respect to unknown individuals as long as the number of these individuals is limited.

The approach is faithful w.r.t. the base formalisms and decidable under the imposed restrictions. It also is tight as the flow of information is possible in both directions. Similar to $\mathcal{DL}+$log, the approach is not flexible as open predicates are interpreted under the open world assumption.

6   NoHR - Querying Ontologies and Rules

In this section, we discuss NoHR[7] (Nova Hybrid Reasoner), a tool for querying combinations of DL ontologies and nonmonotonic rules. It is based on the well-founded semantics for Hybrid MKNF knowledge bases [48], due to its lower computational complexity and amenability to top-down querying without computing the entire model, which is important in the face of huge amounts of data. Indeed, rather than computing the well-founded model for this integration along the ideas presented in Sect. 3.2, queries are evaluated based on $\mathbf{SLG}(\mathcal{O})$, as defined in [2]. This procedure extends SLG resolution with tabling [20] with an *oracle* to $\mathcal{O}$ that handles ground queries to the DL-part of $\mathcal{K}$ by returning (possibly empty) sets of atoms that, together with $\mathcal{O}$ and information already proven true, allows us to derive the queried atom. We refer to [2] for the full account of $\mathbf{SLG}(\mathcal{O})$ and present the idea in the following example.

Example 10. Consider again Example 5 and the query PartialInspection($\mathbf{x}$). Then, the query procedure would find the rule whose head matches the queried atom and proceed by querying for the respective body elements ShpmtCommod($\mathbf{x},\mathbf{y}$) and **not** LowRiskEUCommodity($\mathbf{y}$). We find ShpmtCommod(s_1, c_1), then the remaining query is **not** LowRiskEUCommodity(c_1), which is verified with a query LowRiskEUCommodity(c_1). Now, LowRiskEUCommodity is a DL-predicate and this can be handled by the oracle to $\mathcal{O}$. In fact, LowRiskEUCommodity(c_1) can be inferred from $\mathcal{O}$ if we find ExpeditableImporter($c_1,\mathbf{x}$) and CommodCountry($c_1,\mathbf{y}$) and EUCountry($\mathbf{y}$). Querying for ExpeditableImporter($c_1,\mathbf{x}$) in particular results in a query AdmissibleImporter(i_1) which fails as i_1 is a registered transgressor. Therefore, LowRiskEUCommodity(c_1) eventually fails, and one answer for the initial query is a partial inspection is required for s_1.

While this idea works from a semantic point of view, it leaves open the question of efficiency, as in general there are exponentially many such sets of atoms that together with the ontology allow us to derive the queried atom. The solution applied in NoHR is to transform relevant axioms/logical consequences of the considered ontology into rules, and then take advantage of a rule reasoner to ensure that only polynomially many answers are returned. It has been shown that this process preserves the semantics for the different permitted ontology fragments [21,43,55]. In addition, due to this reasoning approach, DL safety can be relaxed to rule safety, as DL atoms now refer to the rules resulting from the translation.

On the technical side, NoHR[8] is developed as a plug-in for the ontology editor Protégé 5.X,[9] – in fact, the first hybrid reasoner of its kind for Protégé – but it is also available as a library, allowing for its integration within other environments and applications. It supports ontologies written in any of the three

---

[7] http://nohr.di.fct.unl.pt.

[8] The source code can be obtained at https://github.com/NoHRReasoner/NoHR.

[9] http://protege.stanford.edu.

tractable OWL 2 Profiles, and for those combining the constructors permitted in these profiles. Its implementation combines the capabilities of the DL reasoners ELK [47] (for the OWL 2 EL profile), and HermiT [34] and Konclude [75] (for combinations of constructors of different profiles – for OWL 2 QL and RL no DL reasoner is used, rather direct translations into rules are applied) with the rule engine XSB Prolog[10]. XSB comes with support for a vast number of standard built-in Prolog predicates, including numeric predicates and comparisons, and ensures termination of query answering.

NoHR is also robust w.r.t. inconsistencies between the ontology and the rules, which is important as knowledge from different sources may indeed be contradictory on some parts. While this would commonly render the system useless as anything can be derived from an inconsistent knowledge base, a paraconsistent approach is adopted, in which certain parts may be inconsistent – and querying for them with NoHR will reveal that – but other inferences that are not related to such an inconsistency can be inferred as if the inconsistency was not present (see, e.g., [44] for more details). This proves indeed beneficial as inconsistent data on one shipment should not impact on determining whether to inspect another.

NoHR also provides native support for Relational Databases which is encoded through the concept of mappings.[11] Essentially, mappings are used to create predicates that are populated with the result set obtained from queries to external databases, which also allows one to consult tables from different databases. A mapping for predicate p is a triple $\langle p, db, q \rangle$ where the result set from db for the query q (defined over db) is mapped to the predicate p. The connection to database systems is realized using ODBC drivers, thus allowing the integration of NoHR with all major database management systems.

In the following, we describe the system architecture of the Protégé plugin NoHR v4.0 as shown in Fig. 5 and discuss several features of its implementation.

The input for the plugin consists of an OWL file, a rule file and a mappings file. All three components can be edited in Protégé, using the built-in interface for the ontology and the custom "NoHR Rules" and "NoHR Mappings" tabs, provided by the plugin, for the rule and mapping components. The former (as well as the query panel) comes with a dedicated parser to support the creation of correctly formed rules and queries. The latter allows the creation of mappings based on the user's specification of what columns from which tables of which database should be combined, where the underlying SQL queries are dynamically generated, based on the structure of the schema, which allows the automatic application of several optimizations to the generated queries. Alternatively, arbitrary SQL queries can be written to take advantage of the capabilities of the specific DBMS at hand for the sake of, e.g., benefiting from using advanced joins and the associated performance gains when querying.

After the inputs (which can be empty) and the first query are provided, the ontology is translated into a set of rules, using one of the provided reasoners,

---

[10] http://xsb.sourceforge.net.

[11] Similar concepts have been used before for adding database support to rule systems, such as DLV^{DB} [78], and in ontology based data access, such as in ontop [19].

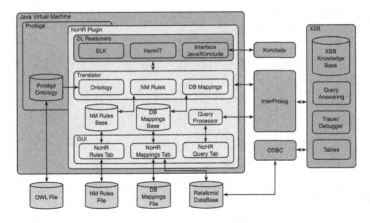

Fig. 5. System architecture of NoHR v4.0 with native database support

ELK [47], HermiT [34] or Konclude [75], depending on the DL in which the ontology is written. The resulting set is then combined with the rules and mappings provided by the input. This joined result serves as input to XSB Prolog via Inter-Prolog,[12] which is an open-source Java front-end, allowing the communication between Java and a Prolog engine, and the query is sent via the same interface to XSB to be executed. During the execution, mappings are providing facts from the external databases as they are requested in the reasoning process. This procedure is supported by the installed ODBC connections and handled within XSB, thus providing full control over the database access during querying and taking advantage of the built-in optimization to access only the relevant part of the database. Answers are returned to the query processor, which displays them to the user in a table (in the Query Tab). Figure 6 provides an example for the query FullInspection(?X), where we note that variables are denoted with a leading "?" to facilitate distinguishing them from constants, similar to SPARQL. The user may pose further queries, and the system will simply send them directly to XSB, without any repeated preprocessing. If the knowledge base is edited, the system recompiles only the component that was changed.

Different versions of NoHR have been evaluated focussing on different aspects [21,43,45,55], and the main observations are summarized in the following.

Different ontologies can be preprocessed for querying in a short amount of time (around one minute for SNOMED CT with over 300,000 concepts), and increasing the number of rules only raises the time for translation linearly. As the preprocessing commonly only needs to be done once before queryin, this is less important for the overall performance. In terms of performance of the different used reasoners for preprocessing, it has been shown [55] that ELK is indeed always fastest, so whenever the ontology fits $\mathcal{EL}_\perp^+$, the dedicated translation module should be used. In between the to general purpose reasoners, HermiT is faster than Konclude on all instances where it does not time out when classifying

---

[12] http://interprolog.com/java-bridge/.

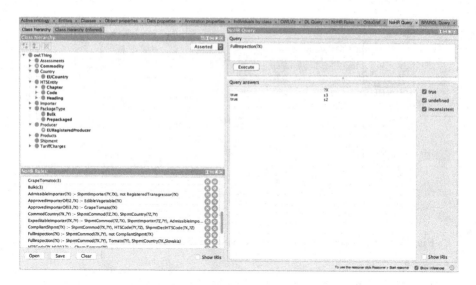

Fig. 6. Cargo assessment example in the NoHR Protégé plugin

the given ontology. Konclude then is preferred in the cases where HermiT fails to classify an ontology that does not fit the OWL 2 EL profile (in which ELK cannot be used either).

In comparison to preprocessing times, querying time is in general neglectable. It has been shown that NoHR scales reasonably well for query answering without non-monotonic rules (only slowing down for memory-intensive cases), even for over a million facts/assertions in the ABox, despite being slightly slower on average for OWL QL in comparison to to the other cases, as part of the OWL inferences is encoded in the rule translations directly, and adding rules scales linearly for pre-processing and querying, even for an ontology with many negative inclusions (such as $DL\text{-}Lite_R$).

In addition, with respect to the database component, it has been shown that, if the data is stored in a database and accessed directly during querying instead of being loaded into memory in the form of facts or ontology assertions, preprocessing time and memory consumption substantially reduces, in particular for tuples of higher arity. In terms of querying, on average querying becomes slightly slower, as the connection via ODBC adds an overhead to the query process. However, if advanced mappings are used, which allow outsourcing certain joins over data from XSB to the DBMS, then improvements of considerable margin can be achieved, in particular when advanced database joins reduce the amount of data that needs to be sent to XSB for reasoning.

7   Conclusions

In this course, we have provided an overview on the integration of Description Logic ontologies and nonmonotonic rules. For that purpose, we have first recalled the two formalisms and reviewed in detail their characteristics and inherent differences. We have then discussed main criteria based on which such an integration can be achieved and argued that often the right choice depends on the application at hand. To illustrate these ideas, we have presented four concrete approaches and compared them with the help of these established criteria. We note that many existing approaches were left out from our presentation. We refer the interested reader to the references mentioned so far (e.g., [27,62] provide detailed discussions of related work), as well as the material from previous Reasoning Web lectures with a different focus [25,50]. We have complemented our considerations on different approaches for such an integration with a more detailed description of one of the reasoning tools, NoHR, that comes with support for databases, robustness to inconsistencies, and fast interactive response times (after a brief preprocessing period), even for larger knowledge bases.

There exists a lot of related work that rather than combining both formalisms aims at combining open and closed world reasoning, by extending one of the two base formalisms with some features from the other. Namely, there is a lot of work on enriching DLs with nonmonotonic reasoning. Description Logics have been extended with default logic [7], with modal operators [23,49] similar to those used in the rules of Hybrid MKNF KBs, circumscription [13,51], as well as defeasible logics [16], and rational closure [33]. On the other hand, rules have been extended with existentials in the head, resulting in Datalog^{+-} [17]. While such rule are undecidable in general, a plethora of different restricted such rule fragments has been defined (see, e.g., [8]) allowing to cover a considerable part of the OWL 2 profiles, and for which answer sets semantics [58] and well-founded semantics [35] have been defined.

For future work, considering dynamics in such combinations of DL ontologies and nonmonotonic rules building on previous work [72–74], in particular in the presence of streams [11] and possibly incorporating heterogeneous knowledge [15,22] seems promising given the huge amounts of data and knowledge that are being created with ever increasing speed and in a variety of formats, for which knowledge-intensive applications are desirable that take advantage of all that information. This certainly is an ambitious objective, but interesting nonetheless.

Acknowledgement. The author thanks Ricardo Gonçalves and the anonymous reviewers for helpful feedback and acknowledges partial support by FCT projects RIVER (PTDC/CCI-COM/30952/2017) and NOVA LINCS (UIDB/04516/2020).

References

1. Alberti, M., Knorr, M., Gomes, A.S., Leite, J., Gonçalves, R., Slota, M.: Normative systems require hybrid knowledge bases. In: van der Hoek, W., Padgham, L., Conitzer, V., Winikoff, M. (eds.) Proceedings of AAMAS, pp. 1425–1426. IFAAMAS (2012)
2. Alferes, J.J., Knorr, M., Swift, T.: Query-driven procedures for hybrid MKNF knowledge bases. ACM Trans. Comput. Log. **14**(2), 1–43 (2013)

3. Alviano, M., et al.: The ASP system DLV2. In: Balduccini, M., Janhunen, T. (eds.) LPNMR 2017. LNCS (LNAI), vol. 10377, pp. 215–221. Springer, Cham (2017). https://doi.org/10.1007/978-3-319-61660-5_19

4. Artale, A., Calvanese, D., Kontchakov, R., Zakharyaschev, M.: The *DL-Lite* family and relations. J. Artif. Intell. Res. (JAIR) **36**, 1–69 (2009)

5. Baader, F., Calvanese, D., McGuinness, D.L., Nardi, D., Patel-Schneider, P.F. (eds.): The Description Logic Handbook: Theory, Implementation, and Applications, 3rd edn. Cambridge University Press, Cambridge (2010)

6. Baader, F., Brandt, S., Lutz, C.: Pushing the $\mathcal{EL}$ envelope. In: Kaelbling, L.P., Saffiotti, A. (eds.) Proceedings of IJCAI, pp. 364–369. Professional Book Center (2005)

7. Baader, F., Hollunder, B.: Embedding defaults into terminological representation systems. J. Autom. Reason. **14**, 149–180 (1995)

8. Baget, J., Leclère, M., Mugnier, M., Salvat, E.: On rules with existential variables: Walking the decidability line. Artif. Intell. **175**(9–10), 1620–1654 (2011)

9. Bajraktari, L., Ortiz, M., Simkus, M.: Combining rules and ontologies into clopen knowledge bases. In: McIlraith, S.A., Weinberger, K.Q. (eds.) Proceedings of AAAI, pp. 1728–1735. AAAI Press (2018)

10. Baral, C., Gelfond, M.: Logic programming and knowledge representation. J. Log. Program. **19**(20), 73–148 (1994)

11. Beck, H., Dao-Tran, M., Eiter, T.: LARS: a logic-based framework for analytic reasoning over streams. Artif. Intell. **261**, 16–70 (2018)

12. Berners-Lee, T., Hendler, J., Lassila, O.: The Semantic Web. Scientific American, pp. 96–101, May 2001

13. Bonatti, P.A., Lutz, C., Wolter, F.: The complexity of circumscription in DLs. J. Artif. Intell. Res. (JAIR) **35**, 717–773 (2009)

14. Brachman, R.J., Levesque, H.J.: Knowledge Representation and Reasoning. Elsevier, Amsterdam (2004)

15. Brewka, G., Ellmauthaler, S., Gonçalves, R., Knorr, M., Leite, J., Pührer, J.: Reactive multi-context systems: heterogeneous reasoning in dynamic environments. Artif. Intell. **256**, 68–104 (2018)

16. Britz, K., Casini, G., Meyer, T., Moodley, K., Sattler, U., Varzinczak, I.: Principles of KLM-style defeasible description logics. ACM Trans. Comput. Log. **22**(1), 1:1-1:46 (2021)

17. Calì, A., Gottlob, G., Lukasiewicz, T., Pieris, A.: Datalog+/-: a family of languages for ontology querying. In: de Moor, O., Gottlob, G., Furche, T., Sellers, A. (eds.) Datalog 2.0 2010. LNCS, vol. 6702, pp. 351–368. Springer, Heidelberg (2011). https://doi.org/10.1007/978-3-642-24206-9_20

18. Calimeri, F., et al.: Asp-core-2 input language format. Theory Pract. Log. Program. **20**(2), 294–309 (2020)

19. Calvanese, D., Cogrel, B., Komla-Ebri, S., Kontchakov, R., Lanti, D., Rezk, M., Rodriguez-Muro, M., Xiao, G.: Ontop: answering SPARQL queries over relational databases. Semantic Web **8**(3), 471–487 (2017)

20. Chen, W., Warren, D.S.: Tabled evaluation with delaying for general logic programs. J. ACM **43**(1), 20–74 (1996)

21. Costa, N., Knorr, M., Leite, J.: Next step for NoHR: OWL 2 QL. In: Arenas, M., et al. (eds.) ISWC 2015. LNCS, vol. 9366, pp. 569–586. Springer, Cham (2015). https://doi.org/10.1007/978-3-319-25007-6_33

22. Dao-Tran, M., Eiter, T.: Streaming multi-context systems. In: IJCAI, pp. 1000–1007. ijcai.org (2017)

23. Donini, F.M., Nardi, D., Rosati, R.: Description logics of minimal knowledge and negation as failure. ACM Trans. Comput. Logic **3**(2), 177–225 (2002)
24. Eiter, T., Fink, M., Ianni, G., Krennwallner, T., Redl, C., Schüller, P.: A model building framework for answer set programming with external computations. TPLP **16**(4), 418–464 (2016)
25. Eiter, T., Ianni, G., Krennwallner, T., Polleres, A.: Rules and ontologies for the semantic web. In: Baroglio, C., et al. (eds.) Reasoning Web. LNCS, vol. 5224, pp. 1–53. Springer, Heidelberg (2008). https://doi.org/10.1007/978-3-540-85658-0_1
26. Eiter, T., Ianni, G., Lukasiewicz, T., Schindlauer, R.: Well-founded semantics for description logic programs in the semantic web. ACM Trans. Comput. Logic. **12**, 11:1-11:41 (2011)
27. Eiter, T., Ianni, G., Lukasiewicz, T., Schindlauer, R., Tompits, H.: Combining answer set programming with description logics for the semantic web. Artif. Intell. **172**(12–13), 1495–1539 (2008)
28. Eiter, T., Šimkus, M.: Linking open-world knowledge bases using nonmonotonic rules. In: Calimeri, F., Ianni, G., Truszczynski, M. (eds.) LPNMR 2015. LNCS (LNAI), vol. 9345, pp. 294–308. Springer, Cham (2015). https://doi.org/10.1007/978-3-319-23264-5_25
29. Gebser, M., Kaufmann, B., Kaminski, R., Ostrowski, M., Schaub, T., Schneider, M.: Potassco: the potsdam answer set solving collection. AI Commun. **24**(2), 107–124 (2011)
30. van Gelder, A.: The alternating fixpoint of logic programs with negation. In: Principles of Database Systems, pp. 1–10. ACM Press (1989)
31. van Gelder, A., Ross, K.A., Schlipf, J.S.: The well-founded semantics for general logic programs. J. ACM **38**(3), 620–650 (1991)
32. Gelfond, M., Lifschitz, V.: Classical negation in logic programs and disjunctive databases. New Gener. Comput. **9**, 365–385 (1991)
33. Giordano, L., Gliozzi, V., Olivetti, N., Pozzato, G.L.: Semantic characterization of rational closure: from propositional logic to description logics. Artif. Intell. **226**, 1–33 (2015)
34. Glimm, B., Horrocks, I., Motik, B., Stoilos, G., Wang, Z.: Hermit: an OWL 2 reasoner. J. Autom. Reason. **53**(3), 245–269 (2014)
35. Gottlob, G., Hernich, A., Kupke, C., Lukasiewicz, T.: Equality-friendly well-founded semantics and applications to description logics. In: AAAI. AAAI Press (2012)
36. Grosof, B.N., Horrocks, I., Volz, R., Decker, S.: Description logic programs: combining logic programs with description logic. In: Hencsey, G., White, B., Chen, Y.R., Kovács, L., Lawrence, S. (eds.) Proceedings of WWW, pp. 48–57. ACM (2003)
37. Haarslev, V., Hidde, K., Möller, R., Wessel, M.: The RacerPro knowledge representation and reasoning system. Seman. Web J. (2011). http://www.semantic-web-journal.net/issues. (to appear)
38. Harris, S., Seaborne, A. (eds.): SPARQL 1.1 Query Language. W3C Working Group Note 21 March 2013 (2013). https://www.w3.org/TR/sparql11-query/
39. Hitzler, P.: A review of the semantic web field. Commun. ACM **64**(2), 76–83 (2021)
40. Hitzler, P., Krötzsch, M., Parsia, B., Patel-Schneider, P.F., Rudolph, S. (eds.): OWL 2 Web Ontology Language: Primer (Second Edition). W3C, Cambridge(2012)
41. Hitzler, P., Krötzsch, M., Rudolph, S.: Foundations of Semantic Web Technologies. Chapman & Hall/CRC, Boca Raton (2009)

42. Horrocks, I., Kutz, O., Sattler, U.: The even more irresistible $\mathcal{SROIQ}$. In: Doherty, P., Mylopoulos, J., Welty, C.A. (eds.) Proceedings of KR, pp. 57–67. AAAI Press (2006)
43. Ivanov, V., Knorr, M., Leite, J.: A query tool for $\mathcal{EL}$ with Non-monotonic Rules. In: Alani, J., et al. (eds.) ISWC 2013. LNCS, vol. 8218, pp. 216–231. Springer, Heidelberg (2013). https://doi.org/10.1007/978-3-642-41335-3_14
44. Kaminski, T., Knorr, M., Leite, J.: Efficient paraconsistent reasoning with ontologies and rules. In: Yang, Q., Wooldridge, M.J. (eds.) Proceedings of IJCAI, pp. 3098–3105. AAAI Press (2015)
45. Kasalica, V., Gerochristos, I., Alferes, J.J., Gomes, A.S., Knorr, M., Leite, J.: Telco network inventory validation with nohr. In: Balduccini, M., Lierler, Y., Woltran, S. (eds.) Logic Programming and Nonmonotonic Reasoning. LPNMR 2019. Lecture Notes in Computer Science, **11481**, pp. 18–331. Springer, Cham (2019). https://doi.org/10.1007/978-3-030-20528-7_2
46. Kazakov, Y.: $\mathcal{RIQ}$ and $\mathcal{SROIQ}$ are harder than $\mathcal{SHOIQ}$. In: Brewka, G., Lang, J. (eds.) Principles of Knowledge Representation and Reasoning: Proceedings of the Eleventh International Conference, KR 2008, Sydney, Australia, September 16–19, 2008. AAAI Press (2008)
47. Kazakov, Y., Krötzsch, M., Simančík, F.: The incredible ELK: from polynomial procedures to efficient reasoning with $\mathcal{EL}$ ontologies. J. Autom. Reason. **53**, 1–61 (2013)
48. Knorr, M., Alferes, J.J., Hitzler, P.: Local closed world reasoning with description logics under the well-founded semantics. Artif. Intell. **175**(9–10), 1528–1554 (2011)
49. Knorr, M., Hitzler, P., Maier, F.: Reconciling OWL and non-monotonic rules for the semantic web. In: Raedt, L.D., Bessiere, C., Dubois, D., Doherty, P., Frasconi, P., Heintz, F., Lucas, P.J.F. (eds.) Proceedings of ECAI. Frontiers in Artificial Intelligence and Applications, vol. 242, pp. 474–479. IOS Press (2012)
50. Krisnadhi, A.A., Maier, F., Hitzler, P.: OWL and rules. In: Reasoning Web 2011, Springer Lecture Notes in Computer Science (2011). http://knoesis.wright.edu/faculty/pascal/resources/publications/OWL-Rules-2011.pdf (to appear)
51. Krisnadhi, A.A., Sengupta, K., Hitzler, P.: Local closed world semantics: Keep it simple, stupid! Technical report, Wright State University (2011). http://pascal-hitzler.de/resources/publications/GC-DLs.pdf
52. Lehmann, J., Isele, R., Jakob, M., Jentzsch, A., Kontokostas, D., Mendes, P.N., Hellmann, S., Morsey, M., van Kleef, P., Auer, S., Bizer, C.: Dbpedia - a large-scale, multilingual knowledge base extracted from Wikipedia. Semantic Web **6**(2), 167–195 (2015)
53. Leone, N., Pfeifer, G., Faber, W., Eiter, T., Gottlob, G., Perri, S., Scarcello, F.: The DLV system for knowledge representation and reasoning. ACM Trans. Comput. Logic **7**, 499–562 (2006)
54. Lifschitz, V.: Nonmonotonic databases and epistemic queries. In: Mylopoulos, J., Reiter, R. (eds.) Proceedings of IJCAI. Morgan Kaufmann (1991)
55. Lopes, C., Knorr, M., Leite, J.: NoHR: integrating XSB Prolog with the OWL 2 profiles and beyond. In: Balduccini, M., Janhunen, T. (eds.) LPNMR 2017. LNCS (LNAI), vol. 10377, pp. 236–249. Springer, Cham (2017). https://doi.org/10.1007/978-3-319-61660-5_22
56. Lukumbuzya, S., Ortiz, M., Simkus, M.: Resilient logic programs: answer set programs challenged by ontologies. In: AAAI, pp. 2917–2924. AAAI Press (2020)
57. Lutz, C.: The complexity of conjunctive query answering in expressive description logics. In: Armando, A., Baumgartner, P., Dowek, G. (eds.) IJCAR 2008. LNCS (LNAI), vol. 5195, pp. 179–193. Springer, Heidelberg (2008). https://doi.org/10.1007/978-3-540-71070-7_16

58. Magka, D., Krötzsch, M., Horrocks, I.: Computing stable models for nonmonotonic existential rules. In: IJCAI, pp. 1031–1038. IJCAI/AAAI (2013)
59. Minker, J., Seipel, D.: Disjunctive logic programming: a survey and assessment. In: Kakas, A.C., Sadri, F. (eds.) Computational Logic: Logic Programming and Beyond. LNCS (LNAI), vol. 2407, pp. 472–511. Springer, Heidelberg (2002). https://doi.org/10.1007/3-540-45628-7_18
60. Morgenstern, L., Welty, C., Boley, H., Hallmark, G. (eds.): RIF Primer (Second Edition). W3C Working Group Note 5 February 2013 (2013). https://www.w3.org/TR/2014/NOTE-rdf11-primer-20140624/
61. Motik, B., Cuenca Grau, B., Horrocks, I., Wu, Z., Fokoue, A., Lutz, C. (eds.): OWL 2 Web Ontology Language: Profiles (Second Edition). W3C, Cambridge (2012)
62. Motik, B., Rosati, R.: Reconciling description logics and rules. J. ACM **57**(5), 93–154 (2010)
63. Motik, B., Sattler, U., Studer, R.: Query-answering for OWL-DL with rules. J. Web Semant. **3**(1), 41–60 (2005)
64. Nenov, Y., Piro, R., Motik, B., Horrocks, I., Wu, Z., Banerjee, J.: RDFox: a highly-scalable RDF store. In: Arenas, M., et al. (eds.) ISWC 2015. LNCS, vol. 9367, pp. 3–20. Springer, Cham (2015). https://doi.org/10.1007/978-3-319-25010-6_1
65. Patel, C., et al.: Matching patient records to clinical trials using ontologies. In: Aberer, K., et al. (eds.) ASWC/ISWC -2007. LNCS, vol. 4825, pp. 816–829. Springer, Heidelberg (2007). https://doi.org/10.1007/978-3-540-76298-0_59
66. Redl, C.: The DLVHEX system for knowledge representation: recent advances (system description). Theory Pract. Log. Program. **16**(5–6), 866–883 (2016)
67. Rosati, R.: On the decidability and complexity of integrating ontologies and rules. J. Web Semant. **3**(1), 41–60 (2005)
68. Rosati, R.: DL+Log: A tight integration of description logics and disjunctive datalog. In: Doherty, P., Mylopoulos, J., Welty, C. (eds.) Tenth International Conference on the Principles of Knowledge Representation and Reasoning, KR 2006, pp. 68–78. AAAI Press (2006)
69. Russell, S.J., Norvig, P.: Artificial Intelligence: A Modern Approach, 4th edn. Pearson, London (2020)
70. Schreiber, G., Raimond, Y. (eds.): RDF 1.1 Primer. W3C Working Group Note 24 June 2014 (2014). available at https://www.w3.org/TR/2014/NOTE-rdf11-primer-20140624/
71. Sirin, E., Parsia, B., Grau, B.C., Kalyanpur, A., Katz, Y.: Pellet: a practical OWL-DL reasoner. Web Semant. **5**, 51–53 (2007)
72. Slota, M., Leite, J.: Towards closed world reasoning in dynamic open worlds. TPLP **10**(4–6), 547–563 (2010)
73. Slota, M., Leite, J., Swift, T.: Splitting and updating hybrid knowledge bases. TPLP **11**(4–5), 801–819 (2011)
74. Slota, M., Leite, J., Swift, T.: On updates of hybrid knowledge bases composed of ontologies and rules. Artif. Intell. **229**, 33–104 (2015)
75. Steigmiller, A., Liebig, T., Glimm, B.: Konclude: system description. J. Web Sem. **27**, 78–85 (2014)
76. Sterling, L., Shapiro, E.: The Art of Prolog - Advanced Programming Techniques, 2nd edn. MIT Press, Cambridge (1994)
77. Swift, T., Warren, D.S.: XSB: extending prolog with tabled logic programming. Theory Pract. Log. Program. **12**(1–2), 157–187 (2012)
78. Terracina, G., Leone, N., Lio, V., Panetta, C.: Experimenting with recursive queries in database and logic programming systems. TPLP **8**(2), 129–165 (2008)

79. Tsarkov, D., Horrocks, I.: FaCT++ description logic reasoner: system description. In: Furbach, U., Shankar, N. (eds.) IJCAR 2006. LNCS (LNAI), vol. 4130, pp. 292–297. Springer, Heidelberg (2006). https://doi.org/10.1007/11814771_26

80. Wang, Y., You, J.H., Yuan, L.Y., Shen, Y.D.: Loop formulas for description logic programs. Theory Pract. Logic Program. **10**(4–6), 531–545 (2010)

81. Xiao, G., et al.: Ontology-based data access: a survey. In: IJCAI, pp. 5511–5519. ijcai.org (2018)

Modelling Symbolic Knowledge Using Neural Representations

Steven Schockaert and Víctor Gutiérrez-Basulto[✉]

Cardiff University, Cardiff, UK
gutierrezbasultov@cardiff.ac.uk

Abstract. Symbolic reasoning and deep learning are two fundamentally different approaches to building AI systems, with complementary strengths and weaknesses. Despite their clear differences, however, the line between these two approaches is increasingly blurry. For instance, the neural language models which are popular in Natural Language Processing are increasingly playing the role of knowledge bases, while neural network learning strategies are being used to learn symbolic knowledge, and to develop strategies for reasoning more flexibly with such knowledge. This blurring of the boundary between symbolic and neural methods offers significant opportunities for developing systems that can combine the flexibility and inductive capabilities of neural networks with the transparency and systematic reasoning abilities of symbolic frameworks. At the same time, there are still many open questions around how such a combination can best be achieved. This paper presents an overview of recent work on the relationship between symbolic knowledge and neural representations, with a focus on the use of neural networks, and vector representations more generally, for encoding knowledge.

1 Introduction

Artificial Intelligence (AI) is built on two fundamentally different traditions, both of which go back to the early days of the field. The first tradition is focused on formalising human reasoning using symbolic representations. This tradition has developed into the Knowledge Representation and Reasoning (KRR) sub-field. The second tradition is focused on learning from examples. This tradition has developed into the Machine Learning (ML) sub-field. These two different traditions have complementary strengths and weaknesses. Due to the use of symbolic representations, KRR systems are explainable, often come with provable guarantees (e.g. on correctness or fairness) and they can readily exploit input from human experts. Moreover, due to their use of systematic reasoning processes, KRR systems are able to derive conclusions that require combining numerous pieces of knowledge in intricate ways. However, symbolic reasoning is too rigid for many applications, where predictions may need to be made about new situations that are not yet covered in a given knowledge base. On the other hand, ML systems often require little human input, but lack explainability, usually come without guarantees, and tend to struggle in applications where systematic

© Springer Nature Switzerland AG 2022
M. Šimkus and I. Varzinczak (Eds.): Reasoning Web 2021, LNCS 13100, pp. 59–75, 2022.
https://doi.org/10.1007/978-3-030-95481-9_3

reasoning is needed [1–3]. Accordingly, there is a growing realisation that future AI systems will need to rely on an integration of ideas from ML and from KRR.

The integration of symbolic reasoning with neural models already has a long tradition within the context of neuro-symbolic AI [4,5]. However, our main focus in this overview is not on the integration of symbolic reasoning with neural network learning, but on the ability of neural network models, and vector space encodings more generally, to play the role of knowledge bases. First, in Sect. 2, we focus on the use of neural models for capturing knowledge graphs (i.e. sets of dyadic relational facts). Knowledge graphs play an important role in research fields such as Natural Language Processing, Recommendation and Machine Learning, essentially giving AI system access to factual world knowledge. The interest in studying the relationship between neural models and knowledge graphs is two-fold. On the one hand, learning vector representations of knowledge graphs makes it easier to use these resources in downstream tasks. On the other hand, existing pre-trained neural language models, trained from large text collections, implicitly capture a lot of the information that is stored in open-domain knowledge graphs. Neural models can thus also play an important role in constructing or extending knowledge graphs. In Sect. 3, we then look at the ability of neural models to capture rules, e.g. the kind of knowledge that would normally be encoded in ontologies. Studying this ability is important because it can suggest mechanisms to combine traditional strategies for rule-based reasoning with neural network learning. Moreover, large pre-trained neural language models can also be used as a source of ontological knowledge, at least to a certain extent. Finally, in Sect. 4 we look at cases where neural models and symbolic knowledge are jointly needed. This includes, for instance, the use of existing rule bases, along with traditional labelled examples, for training neural models. Moreover, symbolic representations are also used for querying neural representations. As a final example, we look at mechanisms to exploit neural representations for making symbolic reasoning more flexible or robust.

2   Encoding Knowledge Graphs

A knowledge graph (KG) is a set of triples of the form $(h, r, t) \in \mathcal{E} \times \mathcal{R} \times \mathcal{E}$, with $\mathcal{E}$ a set of entities and $\mathcal{R}$ a set of relations. A triple (h, r, t) intuitively expresses that the head entity h and tail entity t are in relation r. For instance, (*Cardiff, capital-of, Wales*) asserts that Cardiff is the capital of Wales. KGs are among the most popular frameworks for encoding factual knowledge. Open-domain KGs such as Wikidata [6], YAGO [7] and DBpedia [8], can be seen as providing a structured counterpart to Wikipedia. Such KGs are commonly used as a source of factual encyclopedic information about the world, for instance to enrich neural network models for Natural Language Processing (NLP) [9]. Commonsense KGs such as ConceptNet [10] and ATOMIC [11] are similarly used as a source of knowledge that may otherwise be difficult to obtain. Furthermore, a large number of domain-specific KGs have been developed, for instance covering the needs of a specific business. We refer to [12,13] for a comprehensive overview about knowledge graphs. Here we focus on neural representations of KGs. The aim of using neural representations is to generalise from the facts that are explicitly asserted in a given KG and to make it easier to take advantage of KGs in downstream tasks. In Sect. 2.1, we first discuss KG embedding (KGE) methods, i.e. strategies for learning vector representations of entities and relations that capture the knowledge encoded in a given KG. Such methods have seen a lot of attention from the research community throughout the last decade, having the advantage of being conceptually elegant and computationally efficient. In Sect. 2.2 we then discuss the use of Contextualised Language models (CLMs) such as BERT [14] for capturing knowledge graph triples.

2.1 Knowledge Graph Embeddings

Let a knowledge graph $K \subseteq \mathcal{E} \times \mathcal{R} \times \mathcal{E}$ be given. The aim of knowledge graph embedding (KGE) methods is to learn (i) a vector representation $\mathbf{e}$ for each entity e from $\mathcal{E}$, and (ii) the parameters of a scoring function $f_r : \mathcal{E} \times \mathcal{E} \to \mathbb{R}$ for each relation $r \in \mathcal{R}$, such that $f_r(\mathbf{h}, \mathbf{t})$ reflects the plausibility of the triple (h, r, t). The main focus is usually on the task of *link prediction*, i.e. given a head entity h and relation r, predicting the most likely tail entity t that makes (h, r, t) a valid triple. Embeddings are typically real-valued, i.e. $\mathbf{e} \in \mathbb{R}^n$, but other choices have been considered as well, including complex embeddings [15–17], hyperbolic embeddings [18] and hypercomplex embeddings [19]. In most models, the scoring function f_r is parameterised by a vector $\mathbf{r}$ of the same dimensionality as the entity vectors. For example, in the seminal TransE model [20], we have:

$$f_r(h, t) = -d(\mathbf{h} + \mathbf{r}, \mathbf{t})$$

where d is either the Euclidean or Manhattan distance. In other words, relations are viewed as vector translations, and (h, r, t) is considered plausible if applying the translation for r to $\mathbf{h}$ yields a vector that is similar to $\mathbf{t}$. As another popular example, in DistMult [21], the scoring function is defined as follows:

$$f_r(h, t) = \mathbf{h} \odot \mathbf{r} \odot \mathbf{t}$$

where $\odot$ denotes the component-wise product of vectors. To learn the entity vectors and the scoring functions f_r, several loss functions have been considered, which are typically based on the idea that $f_r(h, t)$ should be higher than $f_r(h, t')$ whenever $(h, r, t) \in K$ and $(h, r, t') \notin K$. An important lesson from research on KGE is that the performance of different methods often crucially depends on the chosen loss function, the type of regularisation that is used, how the negative examples (h, r, t') are chosen, and hyper-parameter tuning [22]. This has complicated the empirical comparison of different KGE models, especially given that these models are typically only evaluated on a small set of benchmarks.

Leaving empirical considerations aside, an important question is whether KGE models have any theoretical limitations on the kinds of KGs they can encode. In other words, is it always possible to find entity vectors and scoring functions such that the triples (h, r, t) which are predicted to be valid by the

KGE model are exactly those that are contained in a given KG? Formally, a KGE model is called fully expressive [23] if for any knowledge graph K, we can find entity vectors and parameters for the scoring functions such that $f_r(h,t) > \gamma$ if $(h, r, t) \in K$ and $f_r(h,t) < \gamma$ otherwise, for some constant $\gamma \in \mathbb{R}$. In other words, a fully expressive model is capable of capturing any knowledge graph configuration. It turns out that basic translation based methods such as TransE are not fully expressive (see [23] for details). However, many other methods have been found to be fully expressive [15,23], provided that vectors of sufficiently high dimensionality are used. This also includes BoxE [24], which is translation based but avoids the limitations of other translation based models by using a region based representation.

2.2   Contextualised Language Models as Knowledge Bases

In recent years, the state-of-the-art in NLP has been based on large pre-trained neural language models (LMs) such as BERT [14]. These LMs are essentially deep neural networks that have been pre-trained on large text collections using different forms of self-supervision. The most common pre-training strategy is based on masked language modelling, where the model is trained to predict words from a given input sentence or paragraph that have been masked. Despite the lack of any explicit supervision signal, the resulting LMs have been found to capture a wealth of syntactic and semantic knowledge [25]. Interestingly, these models also capture a lot of factual world knowledge. For instance, [26] found that presenting BERT with an input such as *"Dante was born in <mask>"* leads to the correct prediction (Florence). In fact, it turns out that pre-trained LMs can be used to answer a wide array of questions, without being given acccess to any external knowledge or corpus [27]. Rather than using KG embeddings to provide NLP models with access to knowledge about the world, the focus in recent years has thus shifted towards (i) analysing to what extent pre-trained LMs already capture such knowledge and (ii) fine-tuning LMs to inject additional knowledge. LMs thus provide a neural encoding of factual world knowledge, although the mechanism by which such knowledge is encoded is unclear. Recent work [28] has suggested that the feedforward layers of these LMs contain neurons that encode specific facts. This insight was used in [29] to devise a strategy to update the knowledge encoded by an LM, for instance when a given fact has become outdated. Some approaches have been suggested for incorporating KGs when training LMs [30], which provides more control about the kind of knowledge that is captured by the LM. Other methods focus on using KGs to reason about the output of LMs [31]. LMs have also been used to aid in the task of KG completion. For instance, [32] designs a scoring function for KG triples, which uses BERT for encoding entity descriptions. Most notably, LMs have been used for link prediction in commonsense KGs such as ConceptNet and ATOMIC. The challenge with such KGs is that entities often correspond to phrases, which may only appear in a single triple. The graph structure is thus too sparse for traditional KG completion methods to be successful. Instead, [33] proposes a model in which a contextualised language model is fine-tuned on

KG triples. They show that, after this fine-tuning step, the LM can be used to generate meaningful new triples. Building on this work, [34] shows that focused commonsense knowledge graphs can be generated on the fly, to provide context for a particular task.

3   Encoding Rules

While knowledge graphs are the *de facto* standard for encoding factual knowledge, more expressive frameworks are needed for encoding generic knowledge. In particular, rules continue to play an important role within AI, and an increasingly important role within NLP. For instance, several competitive strategies for knowledge graph completion based on learned rules have been proposed in recent years [35,36], having the advantage of being more transparent than KGE methods, and the potential for capturing more expressive types of inference patterns. Our focus in this overview is on the interaction between rules and neural representations. First, we discuss the use of neural networks for simulating rule based reasoning in Sect. 3.1. Such methods are particularly appealing, because they are able to learn meaningful rules using standard backpropagation, and can be naturally combined with other types of neural models (e.g. to reason about input presented in the form of images). In Sect. 3.2, we then discuss the view that rules can be modelled in terms of qualitative spatial relationships between region-based representations of concepts and relations. Finally, Sect. 3.3 discusses the rule reasoning abilities of contextualised language models.

Before moving to the next sections, we start by briefly introducing rules; for more details, see e.g. [37]. An *atom* α is an expression of the form $R(t_1, \ldots, t_n)$, where R is a *predicate symbol* with *arity* n and terms t_i, i.e. *variables* or *constants*. An *rule* σ is an expression of the form

$$B_1 \wedge \ldots \wedge B_n \rightarrow \exists X_1, \ldots, X_j.H, \tag{1}$$

where $B_1, \ldots B_n$ and H are atoms and X_m for $1 \leq m \leq j$ are variables. We call $X_1, \ldots, X_j$ the *existential variables of* σ. All other variables occurring in σ are universally quantified. We call a rule with no free variables a *ground rule* and a ground rule with an empty body a *fact*. For example, $fathertOf(john, peter)$, $fathertOf(peter, louise)$ are facts expressing that John is the father of Peter and Peter is the father of Louise. As another example, the following rule with a non-empty body and with variables, defines the grandfather relation in terms of the father relation $fatherOf(X, Y) \wedge fatherOf(Y, Z) \rightarrow grandfatherOf(X, Z)$

3.1   Neural Networks for Reasoning with Differentiable Rules

While the discrete nature of classical logic makes it difficult to integrate with neural networks, several authors have explored techniques for encoding differentiable approximations of logical rules. For instance, [38] develops a differentiable approach to rule based reasoning, called Neural Theorem Proving, by replacing the traditional unification mechanism with a form of soft unification, which is

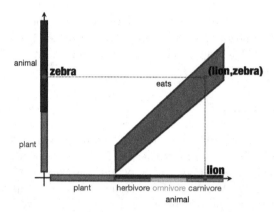

Fig. 1. The binary relation *eats* is defined as a region over the Cartesian product of two conceptual spaces. The spatial configurations capture relational knowledge such as "carnivores only eat animals".

computed based on the dot product between vector representations of the constants and predicates involved. Taking a different approach, Lifted Relational Neural Networks [39] rely on ideas from fuzzy logic to make rules differentiable. In this case, the unification mechanism is the classical one, but truth values of literals and rule bodies are evaluated on a continuous scale. Fuzzy logic connectives are also sometimes used to regularise neural network models based on prior knowledge in the form of rules [40, 41]. DeepProbLog [42] is based on yet another strategy. In this case, reasoning is done using a probabilistic logic program, where a deep neural network is used to estimate the probability of particular literals. Whereas the focus of the aforementioned works was to use neural network learning to discover meaningful rules, in the case of DeepProbLog, the rules themselves are given and the purpose of using neural networks is to allow for more flexible inputs, e.g. making it possible to reason about information presented as images. This strategy has also been instantiated using other logical formalisms, such as answer set programming [43–45]. Some authors have also focused specifically on the use of neural network models for rule induction (rather than for combining rules with neural networks). For instance, [46] presents a differentiable version of inductive logic programming, while [36, 47] propose differentiable models to learn rules for knowledge graph completion.

3.2   Modelling Rules as Spatial Relations

The theory of conceptual spaces [48] was proposed by Gärdenfors as an intermediate representation framework, in between neural and symbolic representations. The main idea is that properties are represented as convex regions, while individuals are represented as points. Compared to the usual vector space models, this region-based approach has the advantage that there is a direct correspondence between spatial relationships in the conceptual space, on the one hand, and symbolic rules, on the other hand. For instance, the fact that individual X satisfies property P, i.e. the fact $P(X)$, corresponds to a situation where the geometric representation of X belongs to the region representing P. Similarly, a rule such as $P(x) \land Q(x) \rightarrow R(x)$ corresponds to the situation where the intersection of the regions representing P and Q is included in the region representing R. While conceptual spaces can only capture propositional knowledge, in [49] we showed how relational knowledge can be similarly modelled by representing relations as convex regions over a Cartesian product of conceptual spaces. Figure 1 illustrates this for the binary relation *eats*, which is defined as a convex region over the Cartesian product of two conceptual spaces. The points in this Cartesian product space correspond to pairs of individuals. Relational knowledge is then modelled in terms of inclusions, intersections and projections. For the example illustrated in the figure, among others, the following rules are captured:

$$carnivore(x) \land eats(x, y) \rightarrow animal(y)$$
$$carnivore(x) \rightarrow \exists y \,.\, animal(y) \land eats(x, y)$$

The framework of conceptual spaces, and their relational extension, seems like a natural choice for settings where neural representations need to be combined with symbolic knowledge. In practice, however, their usage is complicated by the fact that learning regions in high-dimensional spaces is difficult, unless drastically simplifying assumptions are made about the nature of the regions. For example, box embeddings, where entities are represented by hyper-boxes, have been successfully used in a number of contexts [50]. Cones [51,52] and linear subspaces [53] are also common choices. A typical assumption in conceptual spaces is that regions are defined in terms of the prototypes of the corresponding concepts. Region boundaries may then arise as the cells of a (generalised) Voronoi tessellation of these prototypes [54]. This view is appropriate whenever a *contrast set* [55], i.e. a set of jointly exhaustive and pairwise disjoint concepts is given. In [56], an approach was developed for learning concept representations based on this idea.

3.3   Contextualised Language Models as Rule-Based Reasoners

In Sect. 2.2, we discussed how large pre-trained language models encode a substantial amount of factual knowledge. The extent to which such language models capture rules is less clear. In [57], some evidence is provided to suggest that LMs are indeed capable of learning some kinds of symbolic knowledge, and can be trained to apply this knowledge, e.g. generalising an observation about a given concept to hyponyms of that concept. In [58], the ability of transformer based LMs to generalise observed facts is analysed in a systematic way, by training an LM from scratch on a synthetic corpus in which various regularities are present. They find that LMs are indeed capable of discovering symbolic rules, and capable of applying such rules for inferring facts not present in the training corpus, although they also identified important limitations. The aforementioned works mostly focus on one-off rule applications, although some authors have found that

LMs can be trained to perform more sophisticated forms of rule based reasoning [59]. Finally, the ability of transformer based LMs to discover and apply rule-like knowledge has also been exploited in the context of knowledge graph completion. Most notably, [60] shows how a fine-tuned BERT model can essentially be used as a rule base for inductive KG completion.

4   Combining Symbolic Knowledge with Embeddings

In the previous section, we discussed how neural networks are able to capture rule-like knowledge to some extent. In many settings, however, symbolic representations also play a central role. For this reason, we now focus on frameworks for *combining* symbolic and neural representations. For instance, symbolic rules can be used to encode knowledge elicited from a domain expert, hence it is of interest to study mechanisms for incorporating symbolic knowledge when training or using neural models, which we discuss in Sect. 4.1. Symbolic representations are also needed for specifying complex information needs. Recently, approaches have been proposed for evaluating such complex (symbolic) queries against knowledge graph embeddings, and other neural representations (Sect. 4.2). Finally, in applications where interpretability is a primary concern, symbolic knowledge is clearly preferable over neural representations. However, the brittleness of symbolic reasoning means that purely symbolic methods often break down. Symbolic representations are particularly limiting when it comes to inductive reasoning, which in turn makes it difficult to provide plausible or approximate answers in cases where exact reasoning yields no results. To address such concerns, Sect. 4.3 discusses methods in which neural representations are used to add inductive capabilities to symbolic frameworks.

4.1   Injecting Knowledge into Neural Models

Rules are commonly used for injecting prior knowledge when training a neural model [41,61–63]. The most typical strategy is to approximate the rules using differentiable functions, and to add a term to the loss function which encourages the learned representations to adhere to the rules. Another strategy is to use (heuristic) rules to automatically generate (noisy) labelled training examples [64,65]. To train a neural model from these noisy labels, the true label is typically modelled as a latent variable, which is inferred by modelling the reliability of the rules, as well as their correlations in some cases. Rather than using symbolic knowledge during training, some approaches also use symbolic knowledge to reason with the output of a neural model. For instance, [66] proposes a model for question answering, which uses a fine-tuned BERT model to generate a vector representation of the question context (i.e. the question and candidate answer), and then uses a reasoning process which combines that vector with a knowledge graph. The resulting reasoning process uses a Graph Neural Network to dynamically update the question context vector based on the symbolic knowledge captured by the KG. DeepProbLog [42] is also aimed at reasoning about the outputs of a neural network model, in this case based on symbolic probabilistic rules. The general idea of adding a differentiable reasoner on top of a deep neural network model has been explored from a number of other angles. For instance, [67] relies on a differentiable SAT solver to enable reasoning with neural network outputs, while [68] proposes a strategy for using combinatorial optimisation algorithms within an end-to-end differentiable model.

4.2   Complex Query Answering

Learning knowledge graph embeddings has proven a successful approach to predict missing or unobserved edges in knowledge graphs. However, while dealing with knowledge graphs, one is usually interested in handling complex queries describing complex information in the form of graph patterns rather than simple atomic edge-like queries. Indeed, one of the main benefits of symbolically encoded knowledge graphs is that they support SPARQL or conjunctive queries (CQs) [12,13]. However, symbolically encoded KGs can only be queried for existing facts in the knowledge graph, that is, missing entities or edges cannot be inferred. To address this shortcoming, recently various investigations on the use of knowledge graph embeddings to make predictions about conjunctive queries and extensions thereof on incomplete knowledge graphs have been carried out [69–75]. In this setting, for instance, given an incomplete university knowledge graph, we might want to predict *which students are (likely) attending Math and CS modules that use linear algebra?* Unlike for edge (link) prediction, the query might involve several unobserved edges and entities, effectively making this a more complex problem as there exist a combinatorial number of possible interesting queries, and each of them could be matched to many (unobserved) subgraphs of the input KG. In fact, it is not hard to see that a naive approach to query prediction might be unfeasible in practice [69]. One could first use an edge prediction model on all possible pairs of nodes, and then using the obtained edge likelihood, and then enumerate and score all candidate subgraphs that might satisfy a query. However, this enumeration approach is in the worst-case exponential in the number of existentially quantified variables in the query. As a solution, these works represent KG entities and queries in a joint embedding space. For example, the seminal graph query embedding model (GQE) [69] represents KG entities x and a query q as vectors and then cosine similarity is used to score the plausibility of x being a possible answer to q. Most existing query embedding approaches work compositionally by building the embedding of a query q based on its sub-queries. For example, if the input query q is of the form $q_1 \wedge q_2$, the embedding of q is computed based on the embeddings of q_1 and q_2. A number of these works have concentrated on developing query embeddings that support extensions of conjunctive queries, such as positive existential queries (extending CQs with disjunction) [70] or even existential queries with negation [71]. Recently, [75] proposed a framework for answering positive existential queries using pre-trained link predictors to score the atomic queries composing the input query, which is then evaluated using continuous versions of logical operators and combinatorial search or gradient descent. Importantly, this work shows that state-of-the-art results can be obtained using a simple framework requiring only neural link predictors trained for atomic queries, rather than millions of queries as in previous works. In all the works mentioned so far it is assumed that queries have a unique missing entity (answer variable). To overcome this shortcoming, [73] proposed an approach based on transformers to deal with conjunctive queries with multiple missing entities. Finally, [72] investigates whether some of the existing query embedding models are *logically faithful* in the sense that they behave like symbolic logical inference systems with respect to entailed answers. They show that existing models behave poorly in finding logically entailed answers, and propose a model improving faithfulness without losing generalization capacity.

4.3   Using Embeddings for Flexible Symbolic Reasoning

In applications where interpretability is important, using symbolic representations is often preferable. For this reason, developing rule based classifiers remains an important topic of research [35,76,77]. One important disadvantage, however, is that rule bases are usually incomplete. Indeed, learned rules typically only cover situations that are witnessed (sufficiently frequently) in the training data. Neural network models, on the other hand, have the ability to *interpolate* between such situations, which intuitively allows them to make meaningful predictions across a wider range of situations. When rules are manually provided by a domain expert, beyond toy domains we can usually not expect the resulting rule base to be exhaustive either. To address this concern, a number of methods have been proposed which combine the inductive generalisation abilities of neural models, to allow some form of flexible rule-based reasoning. A standard solution is to use vector representations to implement a form of similarity based reasoning [78,79]. Consider, for instance, the following rule: *strawberry → healthy*, and suppose that our knowledge base says nothing about raspberries. Given a standard word embedding [80], we can find out that *strawberry* and *raspberry* are highly similar. Based on this knowledge, we can infer that *raspberry → healthy* is plausible. However, it is difficult to relate degrees of similarity to the plausibility of the inferred rules in a principled way. For this reason, *interpolation* has been put forward as an alternative to similarity based reasoning [81–83]. The intuition is to start from a minimum of two rules e.g. *raspberry → healthy* and *orange → healthy*. Plausible inferences are then supported by the notion of *conceptual betweenness*: we say that a concept B is between the concepts $A_1, ...A_n$ if properties that hold for all of $A_1, ..., A_n$ are also expected to hold for B. If we know that *raspberry* is between *strawberry* and *orange*, then we can plausibly infer the rule *raspberry → healthy* from the two given ones. This interpolation principle is closely related to the notion of category based induction from cognitive science [84]. While this is a general principle, which can be instantiated in different ways, good results have been obtained using strategies which infer betweenness relations from word embeddings and related vector representations [82,85].

5   Concluding Remarks

While it seems clear that future AI systems will somehow need to combine the advantages of symbolic and neural representations, the lack of sufficiently comprehensive symbolic knowledge bases, especially those which capture generic and commonsense knowledge, remains an important obstacle. In the last few years, the focus has somewhat shifted from embedding symbolic knowledge bases to learning knowledge about the world by training deep neural language models. The amount of world knowledge captured by the largest models, such as GPT-3 [86], has been particularly surprising. While by no means perfect, even the commonsense reasoning abilities of these models surpasses expectations[1]. To deal with aspects of commonsense knowledge that are rarely stated in text, a typical strategy in recent years has been to crowdsource targeted natural language assertions and explanations [11,87], and to use such crowdsourced knowledge for fine-tuning language models. However, despite their impressive abilities, neural language models still have two fundamental limitations, which suggest that symbolic representations and systematic reasoning will still play an important role in future AI systems. First, while current NLP models achieve strong results in tasks such as question answering, it is difficult to differentiate between cases where they "know" the answer and cases where they are essentially guessing. Indeed, recent analysis has suggested that language models are still relying on rather shallow heuristics for answering questions [88], which tend to perform well on most benchmarks but offer little in terms of guarantees. Along similar lines, neural machine translation systems are prone to "hallucinating" [89], i.e. generating fluent sentences in the target language which are disconnected from the source text. To use of neural models to make critical decisions thus remains problematic. A second limitation of neural models concerns situations where some form of systematic reasoning is needed. While neural language models can be trained to simulate forward chaining in synthetic settings [59], in practice considerable care is needed to extract the most relevant premises and presenting these in a suitable way, a problem which is studied under the umbrella of multi-hop question answering [90]. Moreover, further progress will need NLP systems to carry out forms of reasoning that go beyond forward or backward chaining, including reasoning about disjunctive knowledge (e.g. arising from the ambiguity of language) and reasoning about the beliefs and intentions of the different participants of a story. It seems unlikely that neural models will be able to carry out such forms for reasoning without relying on some kind of systematic process and structured representation. In fact, for answering questions which require commonsense reasoning, some authors have already found that language models can be improved by repeatedly querying them in a systematic way to extract relevant background knowledge, before trying to answer the question [34,91].

---

[1] https://cs.nyu.edu/~davise/papers/GPT3CompleteTests.html.

References

1. Lake, B.M., Baroni, M.: Generalization without systematicity: on the compositional skills of sequence-to-sequence recurrent networks. In: Proceedings of the 35th International Conference on Machine Learning, pp. 2879–2888 (2018)
2. Geiger, A., Cases, I., Karttunen, L., Potts, C.: Posing fair generalization tasks for natural language inference. In: Proceedings of the 2019 Conference on Empirical Methods in Natural Language Processing and the 9th International Joint Conference on Natural Language Processing, pp. 4484–4494 (2019)
3. Sinha, K., Sodhani, S., Dong, J., Pineau, J., Hamilton, W.L.: CLUTRR: a diagnostic benchmark for inductive reasoning from text. In: Proceedings of the 2019 Conference on Empirical Methods in Natural Language Processing and the 9th International Joint Conference on Natural Language Processing, pp. 4505–4514 (2019)
4. d'Avila Garcez, A.S., Broda, K., Gabbay, D.M.: Neural-Symbolic Learning Systems - Foundations and Applications. Perspectives in Neural Computing, Springer, Heidelberg (2002)
5. d'Avila Garcez, A.S., Gori, M., Lamb, L.C., Serafini, L., Spranger, M., Tran, S.N.: Neural-symbolic computing: an effective methodology for principled integration of machine learning and reasoning. FLAP 6(4), 611–632 (2019)
6. Vrandečić, D., Krötzsch, M.: Wikidata: a free collaborative knowledgebase. Commun. ACM 57(10), 78–85 (2014)
7. Rebele, T., Suchanek, F., Hoffart, J., Biega, J., Kuzey, E., Weikum, G.: YAGO: A multilingual knowledge base from Wikipedia, wordnet, and geonames. In: International Semantic Web Conference, pp. 177–185 (2016)
8. Bizer, C., et al.: DBpedia - a crystallization point for the web of data. J. Web Semant. 7(3), 154–165 (2009)
9. IV, R.L.L., Liu, N.F., Peters, M.E., Gardner, M., Singh, S.: Barack's wife Hillary: Using knowledge graphs for fact-aware language modeling. In: Proceedings of the 57th Conference of the Association for Computational Linguistics, pp. 5962–5971 (2019)
10. Speer, R., Chin, J., Havasi, C.: ConceptNet 5.5: an open multilingual graph of general knowledge. In: Proceedings of the AAAI Conference on Artificial Intelligence (2017)
11. Sap, M., et al.: ATOMIC: an atlas of machine commonsense for if-then reasoning. In: Proceedings of the AAAI Conference on Artificial Intelligence, pp. 3027–3035 (2019)
12. Hogan, A., et al.: Knowledge graphs. CoRR abs/2003.02320 (2020)
13. Hogan, A.: Knowledge graphs: research directions. In: Manna, M., Pieris, A. (eds.) Reasoning Web 2020. LNCS, vol. 12258, pp. 223–253. Springer, Cham (2020). https://doi.org/10.1007/978-3-030-60067-9_8
14. Devlin, J., Chang, M., Lee, K., Toutanova, K.: BERT: pre-training of deep bidirectional transformers for language understanding. In: Proceedings of the 2019 Conference of the North American Chapter of the Association for Computational Linguistics: Human Language Technologies, pp. 4171–4186 (2019)
15. Trouillon, T., Dance, C.R., Gaussier, É., Welbl, J., Riedel, S., Bouchard, G.: Knowledge graph completion via complex tensor factorization. J. Mach. Learn. Res. 18, 130:1-130:38 (2017)
16. Sun, Z., Deng, Z., Nie, J., Tang, J.: RotatE: knowledge graph embedding by relational rotation in complex space. In: 7th International Conference on Learning Representations (2019)

17. Garg, D., Ikbal, S., Srivastava, S.K., Vishwakarma, H., Karanam, H., Subramaniam, L.V.: Quantum embedding of knowledge for reasoning. In: Wallach, H., Larochelle, H., Beygelzimer, A., d'Alché-Buc, F., Fox, E., Garnett, R., eds.: Advances in Neural Information Processing Systems (2019)
18. Balazevic, I., Allen, C., Hospedales, T.: Multi-relational poincaré graph embeddings. Adv. Neural Inf. Process. Syst. **32**, 4463–4473 (2019)
19. Zhang, S., Tay, Y., Yao, L., Liu, Q.: Quaternion knowledge graph embeddings. In: Advances in Neural Information Processing Systems, pp. 2731–2741 (2019)
20. Bordes, A., Usunier, N., García-Durán, A., Weston, J., Yakhnenko, O.: Translating embeddings for modeling multi-relational data. In: Advances in Neural Information Processing Systems, pp. 2787–2795 (2013)
21. Yang, B., Yih, W., He, X., Gao, J., Deng, L.: Embedding entities and relations for learning and inference in knowledge bases. In: Proceedings of the 3rd International Conference on Learning Representations (2015)
22. Jain, P., Rathi, S., Mausam, Chakrabarti, S.: Knowledge base completion: Baseline strikes back (again) CoRR abs/2005.00804 (2020)
23. Kazemi, S.M., Poole, D.: SimplE embedding for link prediction in knowledge graphs. In: Proceedings of the 32nd International Conference on Neural Information Processing Systems, pp. 4289–4300 (2018)
24. Abboud, R., Ceylan, İ.İ., Lukasiewicz, T., Salvatori, T.: BoxE: a box embedding model for knowledge base completion. In: NeurIPS (2020)
25. Rogers, A., Kovaleva, O., Rumshisky, A.: A primer in bertology: what we know about how BERT works. Trans. Assoc. Comput. Linguist. **8**, 842–866 (2020)
26. Petroni, F., et al.: Language models as knowledge bases? In: Proceedings of the 2019 Conference on Empirical Methods in Natural Language Processing and the 9th International Joint Conference on Natural Language Processing (EMNLP-IJCNLP), pp. 2463–2473 (2019)
27. Roberts, A., Raffel, C., Shazeer, N.: How much knowledge can you pack into the parameters of a language model? In: Proceedings of the 2020 Conference on Empirical Methods in Natural Language Processing, pp. 5418–5426 (2020)
28. Geva, M., Schuster, R., Berant, J., Levy, O.: Transformer feed-forward layers are key-value memories (2020). arXiv preprint arXiv:2012.14913
29. Dai, D., Dong, L., Hao, Y., Sui, Z., Wei, F.: Knowledge neurons in pretrained transformers (2021). arXiv preprint arXiv:2104.08696
30. Wang, X., et al.: KEPLER: a unified model for knowledge embedding and pretrained language representation. Trans. Assoc. Comput. Linguist. **9**, 176–194 (2021)
31. Yasunaga, M., Ren, H., Bosselut, A., Liang, P., Leskovec, J.: QA-GNN: reasoning with language models and knowledge graphs for question answering. CoRR abs/2104.06378 (2021)
32. Yao, L., Mao, C., Luo, Y.: KG-BERT: BERT for knowledge graph completion. CoRR abs/1909.03193 (2019)
33. Bosselut, A., Rashkin, H., Sap, M., Malaviya, C., Celikyilmaz, A., Choi, Y.: COMET: commonsense transformers for automatic knowledge graph construction. In: Proceedings of the 57th Annual Meeting of the Association for Computational Linguistics, pp. 4762–4779 (2019)
34. Bosselut, A., Bras, R.L., Choi, Y.: Dynamic neuro-symbolic knowledge graph construction for zero-shot commonsense question answering. In: Proceedings of the Thirty-Fifth AAAI Conference on Artificial Intelligence, pp. 4923–4931 (2021)

35. Meilicke, C., Chekol, M.W., Ruffinelli, D., Stuckenschmidt, H.: Anytime bottom-up rule learning for knowledge graph completion. In: Proceedings of the Twenty-Eighth International Joint Conference on Artificial Intelligence, pp. 3137–3143 (2019)
36. Sadeghian, A., Armandpour, M., Ding, P., Wang, D.Z.: DRUM: end-to-end differentiable rule mining on knowledge graphs. In: Wallach, H.M., et al. (eds.) Proceedings of the Annual Conference on Neural Information Processing Systems, pp. 15321–15331 (2019)
37. Russell, S.J., Norvig, P.: Artificial Intelligence: A Modern Approach, 4th edn. Pearson, London (2020)
38. Rocktäschel, T., Riedel, S.: End-to-end differentiable proving. In: Proceedings of the Annual Conference on Neural Information Processing Systems, pp. 3788–3800 (2017)
39. Sourek, G., Aschenbrenner, V., Zelezný, F., Schockaert, S., Kuzelka, O.: Lifted relational neural networks: efficient learning of latent relational structures. J. Artif. Intell. Res. **62**, 69–100 (2018)
40. Donadello, I., Serafini, L., d'Avila Garcez, A.S.: Logic tensor networks for semantic image interpretation. In: Proceedings of the Twenty-Sixth International Joint Conference on Artificial Intelligence, pp. 1596–1602 (2017)
41. Xu, J., Zhang, Z., Friedman, T., Liang, Y., den Broeck, G.V.: A semantic loss function for deep learning with symbolic knowledge. In: Proceedings of the 35th International Conference on Machine Learning, pp. 5498–5507 (2018)
42. Manhaeve, R., Dumancic, S., Kimmig, A., Demeester, T., Raedt, L.D.: DeepProbLog: neural probabilistic logic programming. In: Proceedings of the Annual Conference on Neural Information Processing Systems, pp. 3753–3763 (2018)
43. Yang, Z., Ishay, A., Lee, J.: NeurASP: embracing neural networks into answer set programming. In: IJCAI, pp. 1755–1762. ijcai.org (2020)
44. Dai, W., Xu, Q., Yu, Y., Zhou, Z.: Bridging machine learning and logical reasoning by abductive learning. In: NeurIPS. (2019) 2811–2822
45. Tsamoura, E., Hospedales, T.M., Michael, L.: Neural-symbolic integration: a compositional perspective. In: AAAI, pp. 5051–5060. AAAI Press (2021)
46. Evans, R., Grefenstette, E.: Learning explanatory rules from noisy data. J. Artif. Intell. Res. **61**, 1–64 (2018)
47. Yang, F., Yang, Z., Cohen, W.W.: Differentiable learning of logical rules for knowledge base reasoning. Proc. Ann. Conf. Neural Inf. Process. Syst. **2017**, 2319–2328 (2017)
48. Gärdenfors, P.: Conceptual Spaces - The Geometry of Thought. MIT Press, Cambridge (2000)
49. Gutiérrez-Basulto, V., Schockaert, S.: From knowledge graph embedding to ontology embedding? An analysis of the compatibility between vector space representations and rules. In: Proceedings of the Sixteenth International Conference on Principles of Knowledge Representation and Reasoning, pp. 379–388 (2018)
50. Patel, D., Dasgupta, S.S., Boratko, M., Li, X., Vilnis, L., McCallum, A.: Representing joint hierarchies with box embeddings. In: Proceedings of the Conference on Automated Knowledge Base Construction (2020)
51. Ganea, O., Bécigneul, G., Hofmann, T.: Hyperbolic entailment cones for learning hierarchical embeddings. In: Proceedings of the International Conference on Machine Learning, pp. 1646–1655 (2018)
52. Özçep, Ö.L., Leemhuis, M., Wolter, D.: Cone semantics for logics with negation. In: Proceedings of the International Joint Conference on Artificial Intelligence, pp. 1820–1826 (2020)

53. Garg, D., Ikbal, S., Srivastava, S.K., Vishwakarma, H., Karanam, H.P., Subramaniam, L.V.: Quantum embedding of knowledge for reasoning. In: Proceedings of the Annual Conference on Neural Information Processing Systems, pp. 5595–5605 (2019)
54. Gärdenfors, P., Williams, M.: Reasoning about categories in conceptual spaces. In: Proceedings of the Seventeenth International Joint Conference on Artificial Intelligence, pp. 385–392 (2001)
55. Goldstone, R.L.: Isolated and interrelated concepts. Memory Cogn. **24**(5), 608–628 (1996)
56. Bouraoui, Z., Camacho-Collados, J., Anke, L.E., Schockaert, S.: Modelling semantic categories using conceptual neighborhood. In: Proceedings of the Thirty-Fourth AAAI Conference on Artificial Intelligence, pp. 7448–7455 (2020)
57. Talmor, A., Tafjord, O., Clark, P., Goldberg, Y., Berant, J.: Leap-of-thought: teaching pre-trained models to systematically reason over implicit knowledge. In: Proceedings of the Annual Conference on Neural Information Processing Systems (2020)
58. Kassner, N., Krojer, B., Schütze, H.: Are pretrained language models symbolic reasoners over knowledge? In: Proceedings of the 24th Conference on Computational Natural Language Learning, pp. 552–564 (2020)
59. Clark, P., Tafjord, O., Richardson, K.: Transformers as soft reasoners over language. In: Proceedings of the Twenty-Ninth International Joint Conference on Artificial Intelligence, pp. 3882–3890(2020)
60. Zha, H., Chen, Z., Yan, X.: Inductive relation prediction by BERT. CoRR abs/2103.07102 (2021)
61. Rocktäschel, T., Singh, S., Riedel, S.: Injecting logical background knowledge into embeddings for relation extraction. In: Proceedings of the Conference of the North American Chapter of the Association for Computational Linguistics: Human Language Technologies, pp. 1119–1129 (2015)
62. Hu, Z., Ma, X., Liu, Z., Hovy, E.H., Xing, E.P.: Harnessing deep neural networks with logic rules. In: Proceedings of the 54th Annual Meeting of the Association for Computational Linguistics (2016)
63. Li, T., Srikumar, V.: Augmenting neural networks with first-order logic. In: Proceedings of the 57th Conference of the Association for Computational Linguistics, pp. 292–302 (2019)
64. Ratner, A., Bach, S.H., Ehrenberg, H., Fries, J., Wu, S., Ré, C.: Snorkel: rapid training data creation with weak supervision. VLDB J. **29**, 709–730 (2019). https://doi.org/10.1007/s00778-019-00552-1
65. Awasthi, A., Ghosh, S., Goyal, R., Sarawagi, S.: Learning from rules generalizing labeled exemplars. In: Proceedings of the 8th International Conference on Learning Representations (2020)
66. Yasunaga, M., Ren, H., Bosselut, A., Liang, P., Leskovec, J.: QA-GNN: reasoning with language models and knowledge graphs for question answering. In: Proceedings of the 2021 Conference of the North American Chapter of the Association for Computational Linguistics: Human Language Technologies, pp. 535–546 (2021)
67. Wang, P.W., Donti, P., Wilder, B., Kolter, Z.: SATNet: bridging deep learning and logical reasoning using a differentiable satisfiability solver. In: Proceedings of the International Conference on Machine Learning, pp. 6545–6554 (2019)
68. Niepert, M., Minervini, P., Franceschi, L.: Implicit MLE: backpropagating through discrete exponential family distributions. CoRR abs/2106.01798 (2021)
69. Hamilton, W.L., Bajaj, P., Zitnik, M., Jurafsky, D., Leskovec, J.: Embedding logical queries on knowledge graphs. In: NeurIPS, pp. 2030–2041 (2018)

70. Ren, H., Hu, W., Leskovec, J.: Query2box: Reasoning over knowledge graphs in vector space using box embeddings. In: ICLR, OpenReview.net (2020)
71. Ren, H., Leskovec, J.: Beta embeddings for multi-hop logical reasoning in knowledge graphs. In: NeurIPS (2020)
72. Sun, H., Arnold, A.O., Bedrax-Weiss, T., Pereira, F., Cohen, W.W.: Faithful embeddings for knowledge base queries. In: NeurIPS (2020)
73. Kotnis, B., Lawrence, C., Niepert, M.: Answering complex queries in knowledge graphs with bidirectional sequence encoders. In: AAAI, pp. 4968–4977. AAAI Press (2021)
74. Choudhary, N., Rao, N., Katariya, S., Subbian, K., Reddy, C.K.: Self-supervised hyperboloid representations from logical queries over knowledge graphs. In: WWW, pp. 1373–1384. ACM/IW3C2 (2021)
75. Arakelyan, E., Daza, D., Minervini, P., Cochez, M.: Complex query answering with neural link predictors. In: Proceedings of the 9th International Conference on Learning Representations (2021)
76. Galárraga, L.A., Teflioudi, C., Hose, K., Suchanek, F.: AMIE: association rule mining under incomplete evidence in ontological knowledge bases. In: Proceedings of the 22nd International Conference on World Wide Web, pp. 413–422 (2013)
77. Omran, P.G., Wang, K., Wang, Z.: Scalable rule learning via learning representation. In: Proceedings of the Twenty-Seventh International Joint Conference on Artificial Intelligence, pp. 2149–2155 (2018)
78. d'Amato, C., Fanizzi, N., Fazzinga, B., Gottlob, G., Lukasiewicz, T.: Ontology-based semantic search on the web and its combination with the power of inductive reasoning. Ann. Math. Artif. Intell. 65(2–3), 83–121 (2012)
79. Beltagy, I., Chau, C., Boleda, G., Garrette, D., Erk, K., Mooney, R.J.: Montague meets Markov: deep semantics with probabilistic logical form. In: Proceedings of the Second Joint Conference on Lexical and Computational Semantics, pp. 11–21 (2013)
80. Mikolov, T., Yih, W., Zweig, G.: Linguistic regularities in continuous space word representations. In: Proceedings NAACL-HLT, pp. 746–751 (2013)
81. Schockaert, S., Prade, H.: Interpolative and extrapolative reasoning in propositional theories using qualitative knowledge about conceptual spaces. Artif. Intell. 202, 86–131 (2013)
82. Derrac, J., Schockaert, S.: Inducing semantic relations from conceptual spaces: a data-driven approach to plausible reasoning. Artif. Intell. 228, 66–94 (2015)
83. Ibáñez-García, Y., Gutiérrez-Basulto, V., Schockaert, S.: Plausible reasoning about el-ontologies using concept interpolation. In: Proceedings of the 17th International Conference on Principles of Knowledge Representation and Reasoning, pp. 506–516 (2020)
84. Osherson, D.N., Smith, E.E., Wilkie, O., Lopez, A., Shafir, E.: Category-based induction. Psychol. Rev. 97(2), 185–200 (1990)
85. Bouraoui, Z., Schockaert, S.: Automated rule base completion as bayesian concept induction. In: The Thirty-Third AAAI Conference on Artificial Intelligence, pp. 6228–6235 (2019)
86. Brown, T.B., et al.: Language models are few-shot learners. In: Larochelle, H., Ranzato, M., Hadsell, R., Balcan, M., Lin, H. (eds.) Proceedings of the Annual Conference on Neural Information Processing Systems (2020)
87. Mostafazadeh, N., et al.: GLUCOSE: generalized and contextualized story explanations. In: Webber, B., Cohn, T., He, Y., Liu, Y., (eds.) Proceedings of the 2020 Conference on Empirical Methods in Natural Language Processing, pp. 4569–4586 (2020)

88. Dufter, P., Kassner, N., Schütze, H.: Static embeddings as efficient knowledge bases? In: Toutanova, K., et al. (eds.) Proceedings of the 2021 Conference of the North American Chapter of the Association for Computational Linguistics: Human Language Technologies, pp. 2353–2363 (2021)
89. Raunak, V., Menezes, A., Junczys-Dowmunt, M.: The curious case of hallucinations in neural machine translation. In: Toutanova, K., et al. (eds.) Proceedings of the 2021 Conference of the North American Chapter of the Association for Computational Linguistics: Human Language Technologies, pp. 1172–1183 (2021)
90. Yang, Z., Qi, P., Zhang, S., Bengio, Y., Cohen, W., Salakhutdinov, R., Manning, C.D.: Hotpotqa: a dataset for diverse, explainable multi-hop question answering. In: Proceedings of the Conference on Empirical Methods in Natural Language Processing, pp. 2369–2380 (2018)
91. Shwartz, V., West, P., Bras, R.L., Bhagavatula, C., Choi, Y.: Unsupervised commonsense question answering with self-talk. In: Webber, B., Cohn, T., He, Y., Liu, Y. (eds.) Proceedings of the Conference on Empirical Methods in Natural Language Processing, pp. 4615–4629 (2020)

Mining the Semantic Web with Machine Learning: Main Issues that Need to Be Known

Claudia d'Amato[✉]

Dipartimento di Informatica – Università degli Studi di Bari Aldo Moro, Bari, Italy
claudia.damato@uniba.it

Abstract. The Semantic Web (SW) is characterized by the availability of a vast amount of semantically annotated data collections. Annotations are provided by exploiting ontologies acting as shared vocabularies. Additionally ontologies are endowed with deductive reasoning capabilities which allow to make explicit knowledge that is formalized implicitly. Along the years a large number of data collections have been developed and interconnected, as testified by the Linked Open Data Cloud. Currently, seminal examples are represented by the numerous Knowledge Graphs (KGs) that have been built, either as enterprise KGs or open KGs, that are freely available. All of them are characterized by very large data volumes, but also incompleteness and noise. These characteristics have made the exploitation of deductive reasoning services less feasible from a practical viewpoint, opening up to alternative solutions, grounded on Machine Learning (ML), for mining knowledge from the vast amount of information available. Actually, ML methods have been exploited in the SW for solving several problems such as link and type prediction, ontology enrichment and completion (both at terminological and assertional level), and concept leaning. Whilst initially symbol-based solutions have been mostly targeted, recently numeric-based approaches are receiving major attention because of the need to scale on the very large data volumes. Nevertheless, data collections in the SW have peculiarities that can hardly be found in other fields. As such the application of ML methods for solving the targeted problems is not straightforward. This paper extends [20], by surveying the most representative symbol-based and numeric-based solutions and related problems, with a special focus on the main issues that need to be considered and solved when ML methods are adopted in the SW field as well as by analyzing the main peculiarities and drawbacks for each solution.

Keywords: Semantic Web · Machine learning · Symbol-based methods · Numeric-based methods

1 Introduction

The Semantic Web (SW) vision has been introduced with the goal of making the Web machine readable [5], by enriching resources with metadata whose formal semantics is defined in OWL[1] ontologies acting as shared vocabularies to be reused. Ontologies are

---

[1] https://www.w3.org/OWL/.

© Springer Nature Switzerland AG 2022
M. Šimkus and I. Varzinczak (Eds.): Reasoning Web 2021, LNCS 13100, pp. 76–93, 2022.
https://doi.org/10.1007/978-3-030-95481-9_4

also empowered with deductive reasoning capabilities which allow for deriving knowledge that is implicitly encoded. While developing this vision, some limitations [35,67] arose: ontology construction resulted in a time consuming task; being strongly decoupled, ontologies and assertions can be out-of-sync, thus resulting in incomplete, noisy and sometimes inconsistent ontologies due to the actual usage of the conceptual vocabulary in the assertions. These limitations became even more evident when pushing on Linked Data [6,66] for enabling the actual creation of the Web of Data and nowadays with the progressive growth of Knowledge Graphs [36]. As a consequence, multiple necessities emerged: reasoning at large scale; managing noise, inconsistencies and incompleteness in the data collections; (semi-)automatizing tasks such as ontology completion, enrichment (both at schema and assertional level), link prediction; exploiting alternative forms of reasoning complementing the deductive approach.

In order to fill some of these gaps, machine learning (ML) methods have been proposed [17]. Problems such as query answering, instance retrieval and link prediction have been regarded as classification problems. Suitable machine learning methods, often inspired by symbol-based solutions in the *Inductive Logic Programming* (ILP) field (aiming at inducing a hypothesised logic program from background knowledge and a collection of examples), have been proposed [16,28,40,46,69]. Most of them are able to cope with the expressive SW representations and the *Open World Assumption* (OWA) typically adopted, differently from the *Closed Wold Assumption* (CWA) that is usually assumed in the traditional ML settings. Problems such as ontology refinement and enrichment at terminology level, e.g. assessing disjointness axioms or complex descriptions for a given concept name, have been regarded as concept learning problems to be solved via supervised/unsupervised inductive learning methods for Description Logic [4] (DLs) representations [24–26,45,62,73].

Nowadays, numeric-based (also called sub-symbolic) ML methods, such as *embeddings* [14,51,56] and *deep learning* [21], are receiving major attention because of their impressive ability to scale when applied to very large data collections. Mostly KG refinement tasks, such as link/type predictions and triple classifications are targeted, with the goal of improving/limiting incompleteness in KGs. Nevertheless, the important gain, in terms of scalability, that numeric-based methods for the SW are obtaining is penalizing: a) the possibility to have interpretable models as a result of a learning process; b) the ability to exploit deductive (and complementary forms of) reasoning capabilities; c) the expressiveness of the SW representations to be considered and the compliance with the OWA.

In the following, the main problems and ML methods that have been developed in the SW are surveyed along with symbol-based (Sect. 2) and numeric-based (Sect. 3) categories, hence the fundamental peculiarities and issues are discussed. Afterwards, considerations concerning the need for solutions that are able to provide human understandable explanations and, towards this direction, to come up with a unified framework integrating both numeric and symbol-based solutions, are reported in Sect. 4. Conclusions are drawn in Sect. 5.

2   Symbol-Based Methods for the Semantic Web

The first efforts in developing ML methods for the SW have been devoted to solve deductive reasoning tasks over ontologies under an inductive perspective. This was motivated by the necessity of offering an alternative way to perform some forms of reasoning when deductive reasoning was not applicable, for instance because of inconsistencies within ontologies, but also for supplying a solution for reasoning in presence of incompleteness (that is when missing information with respect to a certain domain of reference is registered, e.g. missing disjointness axioms), and/or in presence of noise (that is when ontologies are consistent but the information therein is somehow wrong with respect to a reference domain, e.g. missing and/or wrong (derivation of) assertions). Particularly, the incompleteness of knowledge bases, both at assertional and schema level, drove the development of ML methods trying to specifically tackle this problem. The overall idea consisted in exploiting the evidence coming from assertional knowledge for drawing plausible conclusions to be possibly represented with intensional models. In the following, the tasks that received major attention are reported jointly with the analysis of the main solutions for them.

2.1   Instance Retrieval

One of the first problems that has been investigated is the *instance retrieval* problem, which amounts to assessing if an individual is an instance of a given concept. It has been regarded as a classification problem aiming at assessing the class-membership of an individual with respect to a query concept. Similarity-based methods, such as *K-Nearest Neighbor* and *Support Vector Machine*, have been developed since they are well known to be noise tolerant [8, 16, 59]. This required to cope with: 1) the OWA rather than the CWA generally adopted in ML; 2) the non-disjointness of the classes (since an individual can be instance of more that one concept at the same time) while, in the usual ML setting, classes are assumed to be disjoint; 3) the definition of new *similarity measures* and *kernel functions* for exploiting the expressiveness of SW representations. Additionally, because of the OWA, new metrics for the evaluation of the classification results have been defined [16]. This is because, by using standard metrics such as precision, recall and F-measure, new inductive results were deemed as mistakes whilst they could turn out to be correct inferences when judged by a knowledge engineer. The new metrics do not have a direct mapping to the sets of true/false positives/negatives, rather because of the OWA, they consider the cases of unknown/unlabeled results. Particularly, *match rate*, *omission error*, *commission error* and *induction rate* have been proposed. The match rate measures the rate of classification results in agreement with the labels (provided by the use of a standard deductive reasoner). The omission rate measures the cases in which the inductive classifier was not able to provide results, due to the abundance of unlabeled instances because of the OWA, whilst actual labels were available. The commission error measures the cases in which the classifier provided results opposite to the true labels (e.g. an individual being instance of the negated query concept). The induction rate measures the cases in which the classifier was able to provide a label whilst it was not available due to OWA. The proposed solutions experimentally proved their ability to perform inductive instance retrieval when compared to a standard deductive reasoner. Additionally, they also proved their ability to induce new knowledge that was not logically derivable[2]. Nevertheless, they were not fully able to work at large scale.

Methods characterized by more interpretable models have also been defined [26, 63]. Inspired by the ILP literature on the induction of decision trees in clausal representation [7], a solution for inducing a *Terminological Decision Tree* (TDT) has been formalized [26]. A TDT is a tree structure, naturally compliant with the OWA, employing: a DL language for representing nodes and inference services as corresponding tests on the nodes. The tree-induction algorithm adopts a classical top-down divide-and-conquer strategy with the use of refinement operators for DL concept descriptions. Once a TDT is induced, similarly to logical decision trees, a definition for the target concept (namely the concept with respect to which classification is to be performed) can be drawn, by exploiting the nodes in the tree structure. This solution showed the interesting ability to provide an interpretable model, but it turned out slightly less effective then similarity-based classification methods.

Nevertheless, when assessing the concept membership for an individual, as recalled above, it may result instance of more than one concept at the same time. As such a more suitable way to regard the problem is as *multi-label classification* task [77], where multiple labels (concepts in the specific case) may be assigned to each instance. Some preliminary research has been presented in [50], focussing on type prediction in RDF data collections where limited information from the available background knowledge is considered. *Multiple-instance learning* (MIL) [11] is also a setting that would need investigation. It deals with the problem of incomplete knowledge concerning labels in training sets, as it happens in SW knowledge bases due to OWA. MIL is a type of supervised learning where training instances are not individually labeled, they are collected in sets of labeled bags. From a collection of labeled bags, the learner tries to either (i) induce a concept that will label individual instances correctly or (ii) learn how to label bags without inducing the concept. It may be fruitfully exploited for discovering correlations among resources and/or emerging concepts.

Other settings that would be useful for coping with the large number of unlabelled instances are *semi-supervised learning* (SSL) [12] and *learning from imbalanced data*. SSL makes use of both labeled and unlabeled instances, during the learning process, for surpassing the classification performance that could be obtained by discarding the unlabeled data (as it would happen in a supervised learning setting). Very few research efforts have been made in this direction. Some initial results have been presented in [52], where a link prediction problem is solved in a transductive learning framework. In learning from unbalanced data [32,48], that is data collections where the labels distribution is not uniform, sampling techniques are usually adopted in order to create a balanced dataset to be successively used for the learning task. *Ensemble methods*, consisting in using multiple learning algorithms to obtain better predictive performance, could be fruitfully adopted, as illustrated in [27,63] where respectively a *boosting* [27] and *bagging* [63] technique is employed.

---

[2] The induced knowledge should be validated by ontology engineerings for the possible further enrichment of ontologies.

2.2 Concept Learning for Ontology Enrichment

With the purpose of enriching ontologies at terminological level, methods for learn-ing concept descriptions for a concept name have been proposed. The problem has been regarded as a supervised concept learning problem aiming at approximating an intensional DLs definition, given a set of individuals of an ontology acting as posi-tive/negative training examples.

Various solutions, e.g. DL-FOIL [24] and CELOE [45] (part of the DL-LEARNER suite[3]), have been formalized. They are mostly grounded on a *separate-and-conquer* (sequential covering) strategy: a new concept description is built by specializing, via suitable *refinement operators*, a partial solution to correctly cover (i.e. decide a consis-tent classification for) as many training instances as possible. Whilst DL-FOIL works under OWA, CELOE works under CWA. Both of them may suffer from ending up in sub-optimal solutions. In order to overcome such issue, DL-FOCL [64], PARCEL [70] and SPACEL [71] have been proposed. DL-FOCL is an optimized version of DL-FOIL, implementing a base greedy covering learner. PARCEL combines top-down and bottom-up refinements in the search space. The learning problem is split into various sub-problems, according to a divide-and-conquer strategy, that are solved by running CELOE. Once the partial solutions are obtained, they are combined in a bottom-up fash-ion. SPACEL extends PARCEL with a symmetrical specialization of a concept descrip-tion.

These solutions proved their ability to learn approximated concept descriptions for a target concept name but relatively small ontological knowledge bases have been con-sidered for the experiments.

2.3 Knowledge Completion

Knowledge completion consists in finding new information at assertional level, that is facts that are missing in a considered knowledge base. This task has become very popular with the development of KGs, that are well known to be incomplete, and it is also strongly related to the link prediction task (see Sect. 3).

One of the most well known systems for knowledge completion of RDF knowl-edge bases is AMIE [28]. Inspired by the literature in *association rule mining* [2] and ILP methods for learning Horn clauses, AMIE aims to mine logic rules from RDF knowledge bases with the final goal of predicting new assertions. AMIE (and its opti-mized version AMIE+ [29]) currently represents the most scalable rule mining system for learning Horn rules on large RDF data collections and is also explicitly tailored to support the OWA. However, it does not exploit any form of deductive reasoning. A related rule mining system, similarly based on a level-wise generate and test strategy has been proposed in [19]. It aims to learn SWRL rules [37] from OWL ontologies while exploiting schema level information and deductive reasoning during the rule learning process. Both AMIE and the solution presented in [19] showed the ability to mine useful rules and to predict new assertional knowledge. The solution proposed in [19] showed reduced scalability due to the exploitation of the reasoning capabilities.

---

[3] https://dl-learner.org/.

2.4 Learning Disjointness Axioms

Disjointness axioms are essential for making explicit the negative knowledge about a domain, yet they are often overlooked during the modeling process (thus affecting the efficacy of reasoning services). To tackle this problem, automated methods for discovering these axioms from the data distribution have been devised.

A solution grounded on *association rule mining* [2] has been proposed in [72,73]. It is based on studying the correlation between classes comparatively, namely *association rules*, *negative association rules* and *correlation coefficient*. Background knowledge and reasoning capabilities are used to a limited extent.

A different solution has been proposed in [62] where, moving from the assumption that two or more concepts may be mutually disjoint when the sets of their (known) instances do not *overlap*, the problem has been regarded as a clustering problem, aiming at finding partitions of similar individuals of the knowledge base, according to a *cohesion* criterion quantifying the degree of homogeneity of the individuals in an element of the partition. Specifically, the problem has been cast as a *conceptual clustering* problem, where the goal is both to find the best possible partitioning of the individuals and also to induce intensional definitions of the corresponding classes expressed in the standard representation languages. Emerging disjointness axioms are captured by the employment of *terminological cluster trees* (TCTs) and by minimizing the risk of mutual overlap between concepts. Once the TCT is grown, groups of (disjoint) clusters located at sibling nodes identify concepts involved in candidate disjointness axioms to be derived. Unlike [72,73], based on the statistical correlation between instances, the empirical evaluation of [62] showed its ability to discover disjointness axioms also involving complex concept descriptions, thanks to the exploitation of the underlying ontology as background knowledge.

2.5 Capturing Ontology Evolutions

Some acquired knowledge may also evolve over time. For instance, given an ontology, due to the insertion of new individuals and assertions, new concept formations may emerge over time but lacking of intentional definitions (*novelty detection* [68]). Similarly, existing concepts, defined by their intention, may actually evolve towards more general or more specific concepts when looking at they extensions that again may evolve over time (*concept drift* [75]).

Capturing these phenomenon may be fundamental and unsupervised as well as pattern mining methods would be useful for the purpose. Some preliminary research on capturing knowledge evolution by exploiting conceptual clustering methods has been presented [23,25]. Particularly, a two step approach is proposed. As a first step, suitable distance-based and semantically enhanced clustering method are exploited in oder to spot cases of concept that are evolving or novel concepts which are emerging based on the elicited clusters. Afterwards, concept learning algorithms for DL representations (see Sect. 2.2) are used to produce new concepts based on a group of examples (i.e. individuals in a cluster) and counterexamples (individuals in disjoint clusters).

The proposed solutions proved the feasibility of the overall approach by showing the ability to capture new and evolving concepts but also highlighted a main limitation given by the lack of gold standards for validating the results.

3   Numeric-Based Methods for the Semantic Web

Whilst symbolic methods adopt symbols for representing entities and relationships of a domain and infer generalizations that provide new insights into the data and are ideally readily interpretable, numeric-based methods typically adopt feature vector (propositional) representations and cannot provide interpretable models but they are usually rather scalable [49].

The problem that has been mainly investigated in the SW context by adopting numeric solutions is *link prediction* which amounts to predict the existence (or the probability of correctness) of triples in (a portion of) the Web of Data. Data are considered in their graph representation, mostly the RDF representation language has been targeted and almost no reasoning is exploited; most expressive SW languages are basically discarded. The attention towards this problem is also grown due to the increasing availability of KGs, that are known to be often missing facts [74]. In the KG context, link prediction is also referred to as *knowledge graph completion*. Methods borrowed from the Statistical Relational Learning (SRL) [31] (having as main goal the creation of statistical models for relational/graph-based data) have been mostly developed. In the following the main classes of methods and solutions targeting link prediction in the SW are analyzed.

3.1   Probabilistic Latent Variable Models

Probabilistic Latent Variable Models explain relations between entities by associating each resource to a set of intrinsic latent attributes (i.e. attributes not directly observable in the data) and conditions the probability distribution of the relations between two resources on their latent attributes. All relations are considered conditionally independent given the latent attributes. This allows the information to propagate through the network of interconnected latent variables.

One of the first numeric-based link prediction solution belonging to this category is the *Infinite Hidden Semantic Model* (IHSM) [60]. It formalizes a probabilistic latent variable that associates a latent class variable with each resource/node and makes use of constraints expressed in First Order Logic during the learning process. IHSM showed promising results but resulted in a limited scalability on large SW data collection because of the complexity of the probabilistic inference and learning, which is intractable in general [42].

3.2   Embedding Models

With the goal of scaling on very large SW data collections, *embedding models* have been investigated. Similarly to probabilistic latent variable models, in embedding models each resource/node is represented with a continuous embedding vector encoding its intrinsic latent features within the data collection. Models in this class do not necessarily rely on probabilistic inference for learning the optimal embedding vectors and this allows to avoid the issues related to the normalization of probability distributions, that may lead to intractable problems.

Particularly, KG embedding methods have received considerable attention. They typically map entities and relations forming complex graph structures to simpler representations (feature-vectors) and aim at learning prediction functions to be exploited for tasks such as link prediction and triple classification. The scalability purpose motivated the interest delved towards these models [10] which have been shown to ensure good performances on very large KGs. Specifically, KG embedding methods aim at converting the data graph into an optimal low-dimensional space in which *graph structural information* and *graph properties* are preserved as much as possible [10,39]. The low-dimensional spaces enable computationally efficient solutions that scale better with the KG dimensions. Graph embedding methods may differ in their main building blocks: the *representation space* (e.g. point-wise, complex, discrete, Gaussian, manifold), the *encoding model* (e.g. linear, factorization, neural models) and the *scoring function* (that can be based on distance, energy, semantic matching or other criteria) [39]. In any case, the objective consists in learning embeddings such that the score of a valid (positive) triple is lower than the score of an invalid triple standing for a sort of negative examples.

One of the first solutions belonging to this category is RESCAL [56], which implements graph embedding by computing a three-way factorization of an adjacency tensor that represents the multi-graph structure of the data collection. RESCAL resulted in a powerful model, it was also able to capture complex relational patterns over multiple hops in a graph, however, even if improving the scalability of IHSM, it was not able to scale on very large graph-based data collections (e.g. the whole YAGO or DBPedia). The main limitation was represented by the parameter learning phase, which may take rather long for converging to optimal solutions. With the goal of improving the model training phase employed by RESCAL, a solution exploiting adaptive learning rates during training has been proposed [51]. Specifically, an energy-based embedding model has been formalized, where entities and relations are embedded in continuous vector spaces and the probability of an RDF triple to encode a true statement is expressed in terms of energy of the triple, which is an unnormalized score that is inversely proportional to such a probability value. It is computed as a function of the embedding vectors of the subject, the predicate and the object of the triple. This solution experimentally showed improvements in terms of efficiency of the parameter learning process and more accurate results in a significantly lower number of iterations.

Nevertheless, the very first embedding model that registered very high scalability performances has been TRANSE [9]. It introduces a very simple but effective and efficient model: each entity is represented by an embedding vector and each predicate is represented by a (vector) *translation operation*. The score of a triple is given by the similarity (negative L_1 or L_2 distance) of the translated subject embedding to the object embedding. The optimal embedding and translation vectors for predicates are learned jointly. The method relies on a *stochastic optimization process*, that iteratively updates the distributed representations by increasing the score of the positive triples i.e. the observed triples, while lowering the score of unobserved triples standing as negative examples. The embedding of all entities and predicates in the KG is learned by minimizing a *margin-based ranking loss*.

Despite the scalability of TRANSE, it resulted limited in representing properly various types of properties such as *reflexive* ones, and 1-to-N, N-to-1 and N-to-N relations. To tackle this limitation, while keeping the ability to scale on very large KGs, moving from TRANSE, a large family of models has been developed. Among others, TRANSR [47] has been proposed as a more suitable model to handle non 1-to-1 relations. It adopts a score function that preliminarily projects entities and relations to the different spaces and successively they are put together through a suitable projection matrix. The main variation introduced by the new model regards the way the entities are projected in the vector space of the relations, which increases the complexity without compromising the overall scalability.

An important point that needs to be highlighted is that, due to tackling RDF representations, most of the considered data collections only contain positive (training) examples, since usually false facts are not encoded. As training a learning model in all-positive examples could be tricky because the model might easily over generalize, for obtaining negative examples two different approaches are generally adopted: either *corrupting* true/observed triples with the goal of generating plausible negative examples or making a *local-closed world assumption* (LCWA) in which the data collection is assumed as *locally* complete [55]. In both cases, wrong negative information may be generated and thus used when training and learning the embedding models; hence alternative solutions are currently investigated [3]. Even more so, existing embedding models do not make use of the additional semantic information that may be coded when more expressive representation languages are adopted. Indeed the need for *semantic embedding methods* has been argued [20,38,57].

3.3 Semantically Enriched Embedding Models

Recently, semantically empowered embedding models, particularly targeting KG refinement tasks, have been investigated [18,38,53] and various approaches have been proposed that leverage different specific forms of prior knowledge to learn better representations.

In [33] a KG embedding method considering also logical rules has been proposed, where triples in the KG and rules are represented in a unified framework. Specifically, triples are represented as atomic formulae while rules are represented as more complex formulae modeled by t-norm fuzzy logics admitting as antecedent single atoms or conjunctions of atoms with variables as subjects and objects. A common loss over both representation is defined which is minimized to learn the embeddings. This proposal resulted in a novel solution but the specific form of prior knowledge that has to be available for the KG constitutes its main drawback. A similar drawback also applies to the model proposed in [54], where a solution based on adversarial training is formalized, exploiting Datalog clauses to encode assumptions which are used to regularize neural link predictors. An inconsistency loss is derived that measures the degree of violation of such assumptions on a set of adversarial examples. Training is defined as a minimax problem, in which the models are trained by minimizing the inconsistency loss on the adversarial examples jointly with a supervised loss. Nevertheless, in [1] the limitations of the current embedding models have been identified: theoretical inexpressiveness, lack of support for inference patterns, higher-arity relations, and logical rule incorporation.

Complementary solutions, besides exploiting the graph structural information and properties, focused on exploiting also the additional knowledge available, when rich representation languages as RDFS and OWL are employed, that is when no specific addtional formalisms are required for representing additional prior knowledge. Particularly, [53] has proven the effectiveness of combinations of embedding methods and strategies relying on reasoning services for the injection of *Background Knowledge* (BK) to enhance the performance of a specific predictive model. Following this line, TRANSOWL, aiming at injecting BK particularly during the learning process, and its upgraded version TRANSROWL, where a newly defined and more suitable loss function and scoring function are also exploited, have been proposed [18]. The main focus is on the application of this idea to enhance well-known basic scalable models, namely TRANSE [9] and TRANSR [47][4], even if, in principle, the proposed approach could be applied to more complex embedding methods, with an additional formalization. The proposed solutions can take advantage of an informed corruption process that leverages on reasoning capabilities, while limiting the amount of false negatives that a less informed random corruption process may cause.

In TRANSOWL the original TRANSE setting is maintained while resorting to reasoning with schema axioms to derive further triples to be considered for training and that are generated consistently with the semantics of the properties. Particularly, for each considered axiom, TRANSOWL defines, on the score function, specific constraints that guide the way embedding vectors are learned. It extends the approach in [53], formalizing a model characterized by two main components devoted to inject BK in the embedding-based model during the training phase: 1) *Reasoning:* It is used for generating corrupted triples that can certainly represent negative instances, thus avoiding false negatives, for a more effective model training. Moreover, false positives can be detected and avoided. Specifically, using a reasoner[5] it is possible to generate corrupted triples exploiting the available axioms specified in RDFS and OWL. The following axioms are considered: domain, range, disjointWith, functionalProperty; 2) *BK Injection:* A set of different axioms, specifically equivalentClass, equivalentProperty, inverseOf and subClassOf, are employed for the definition of constraints on the score function considered in the training phase so that the resulting vectors, related to such axioms, reflect their specific properties. As a consequence, new triples are also added to the training set on the grounds of the specified axioms.

TRANSROWL further evolves the approach used to derive TRANSOWL from TRANSE by adopting TRANSR as the base model in order to handle non 1-to-1 properties in a more proper way. Indeed, the poor modeling of these relations (caused by TRANSE) may generate spurious embedding vectors with null values or analogous vectors among different entities, thus compromising the ability of making correct predictions. A noteworthy case regards the typeOf property, a common N-to-N relationship. Modeling such property with TRANSE amounts to a simple vector translation; the considered individuals and classes may be quite different in terms of properties and attributes they are involved in, thus determining strong semantic differences (according

---

[4] TRANSR tackles some weak points in TRANSE, such as the difficulty of modeling specific types of relationships [3].

[5] Facilities available in the Apache Jena framework were used: https://jena.apache.org.

to [76]) taking place at large reciprocal distances in the underlying vector space, hence revealing the weakness of employing the mere translation. Differently, TRANSR associates to typeOf, and to all other properties, a specific vector space where entity vectors are projected to. This leads to training specific projection matrices for typeOf so that the projected entities can be located more suitably to be linked by the vector translation associated to typeOf.

These models, characterized by learning embeddings whilst exploiting prior knowledge both during the learning process and the triple corruption process, have been proved to improve their effectiveness compared to the original models, that focus on structural graph properties with a random corruption process, on link prediction and triple classification tasks. Nevertheless, they also showed some shortcomings since they suffered when some of the considered schema axioms were missing, thus suggesting that further research needs to be pursued in this direction.

3.4 Vector Space Embeddings for Propositionalization

A complementary research direction focused on the exploitation of vector space embeddings for obtaining a propositional feature vector representation of RDF data collections. Specifically, inspired by the data mining (DM) literature on propositionalization [43], that is a collection of methods for transforming a relational data representation into a (numeric) propositional feature vector representation so that scalable propositional DM/ML methods can be applied, RDF2Vec [61] has been proposed. It formalizes a solution for learning latent numeric representations of entities in RDF graphs by adapting language modeling approaches. A two-steps approach is adopted: first the RDF graph is converted into a set of sequences of entities (for the purpose two different approaches using local information, that are graph walks and Weisfeiler-Lehman Subtree RDF graph kernels, are exploited); in the second step, the obtained sequences are used to train a neural language model estimating the likelihood of a sequence of entities appearing in a graph. The outcome of the the training process provides each entity in the graph represented as a vector of latent numerical features. DBpedia and Wikidata have been processed. In order to show that the obtained vector representation is independent from task and algorithm, an experimental evaluation involving a number of classification and regression tasks has been performed.

An upgrade of RDF2Vec has been presented in [14]. The proposed solution is grounded on the exploitation of global patterns, differently from RDF2Vec which exploits local patterns. None of the two solutions can cope with literals.

4   On the Need for Explainable Solutions

The need to cope with the fast growing of the Web of Data and the emerging very large KGs required the SW community to show its ability to manage such tremendous amount of data and knowledge.

This mostly motivated the right attention towards numeric ML methods, particularly for providing scalable solutions to manage the inherent incompleteness of the Web of

Data. Indeed, current symbolic methods are not actually comparable, in terms of scalability, to numeric-based solutions. This gain is not for free. It is obtained by giving up the expressive representation languages, such as OWL, that the SW community contributed to standardize with the goal of formalizing rich and expressive knowledge, but also by almost forgetting one of the most powerful characteristic of these languages, that is being empowered with deductive reasoning capabilities that allow for deriving new knowledge. This means to loose knowledge that is already available. Indeed, as illustrated in Sect. 3, almost all numeric methods focus on RDF as a representation language and nearly no reasoning capabilities are exploited. Furthermore, differently from symbolic methods, numeric-based solutions lack the ability to provide interpretable models (see Sect. 3), thus limiting the possibility to interpret and understand the motivations for the returned results. Additionally, tasks such as learning concept or disjointness axioms cannot be performed without symbol-based methods which can certainly benefit of the very large amount of information to provide potentially more accurate results.

Research efforts need to be devoted towards ML solutions that, while keeping scalability, are able to target more expressive representations as well as to provide interpretable models. As a first step, the integration of numeric and symbolic approaches should be focused on.

Some discussions in this direction have been developed by the Neural-Symbolic Learning and Reasoning community [30,34], which seeks to integrate principles from neural networks learning and logical reasoning. The main conclusion has been that neural-symbolic integration appears particularly suitable for applications characterized by the joint availability of large amounts of (heterogeneous) data and knowledge descriptions, which is actually the case of the Web of Data. A set of key challenges and opportunities have been outlined [30], such as: how to represent expressive logics within neural networks, how neural networks should reason with variables, or how to extract symbolic representation from trained neural networks. Preliminary results for some of these challenges have been recently registered, encouraging pursuing the research direction. An example is represented by SimplE [41], a scalable tensor-based factorization model that is able to learn interpretable embeddings incorporating logical rules through weight tying. Ideas for extracting propositional rules from trained neural networks under SW background knowledge have been illustrated [44], showing that the exploitation of BK allows for: reducing the extracted rule set; reproducing the input-output function of the trained neural network. A conceptual sketch for explaining the classification behavior of artificial neural networks in a non-propositional setting while using SW background knowledge has been proposed [65]. This sheds the light on another important issue, that is the necessity to provide explanations for results supplied by ML methods [15], particularly when they come from very large sources of knowledge, e.g. results for a link prediction problem.

The solution depicted in [65] is in agreement with the idea of exploiting symbol-based interpretable models to explain conclusions [49,58]. Nevertheless, interpretable models describe *how* solutions are obtained but not *why* they are obtained. As argued in [22,30], providing an explanation means to supply a line of reasoning, illustrating the decision making process of a model whilst using human understandable features. Following this direction, a solution providing human-centric transfer learning explanation has been proposed [13]. It takes advantage of ontologies (DBPedia is used) and reasoning capabilities to infer different kinds of human understandable explanatory evidence. Hence, in a more broad sense, providing an explanation means to open the box of the reasoning process and make it understandable. In a complex setting such as the Web of Data, where knowledge may result from an automatic information acquisition and integration process from different sources, thus potentially noisy and with conflicting information, multiple reasoning paradigms may be required e.g. deduction (when rules and theory are available), induction (for building models from the available knowledge), abduction (for filling in partial models coping with incomplete theory), commonsense reasoning etc. Large research efforts have been devoted to study each paradigm, however in the considered complex scenario, multiple paradigms could be needed at the same time. This may require the formalization of a unifying reasoning framework.

5   Conclusions

This paper surveyed the different SW problems where ML solutions have been employed and the progresses that have been registered from them. A major focus has been devoted to the main issues that need to be considered and solved when ML solutions are adopted in the SW field. Specifically, symbol-based and numeric-based methods have been analyzed and their main peculiarities and drawbacks have been highlighted. Hence, some considerations concerning the need for solutions that are able to provide human understandable explanations and, towards this direction, to come up with a unified framework integrating both numeric and symbol-based solutions, have been reported.

References

1. Abboud, R., Ceylan, İ.İ., Lukasiewicz, T., Salvatori, T.: BoxE: a box embedding model for knowledge base completion. In: Proceedings of NeurIPS 2020 (2020)
2. Agrawal, R., Imieliński, T., Swami, A.: Mining association rules between sets of items in large databases. In: Buneman, P., et al. (eds.) Proceedings of the 1993 ACM SIGMOD International Conference on Management of Data, pp. 207–216. ACM Press (1993). https://doi.org/10.1145/170035.170072
3. Arnaout, H., Razniewski, S., Weikum, G.: Enriching knowledge bases with interesting negative statements. In: Das, D., et al. (eds.) Proceedings of AKBC 2020 (2020). https://doi.org/10.24432/C5101K
4. Baader, F., Calvanese, D., McGuinness, D.L., Nardi, D., Patel-Schneider, P.F. (eds.): Description Logic Handbook, 2nd edn. Cambridge University Press, Cambridge (2010). https://doi.org/10.1017/CBO9780511711787
5. Berners-Lee, T., Hendler, J., Lassila, O.: The semantic web. Sci. Am. **284**(5), 34–43 (2001). https://doi.org/10.4018/jswis.2009081901
6. Bizer, C., Heath, T., Berners-Lee, T.: Linked data - the story so far. Int. J. Sem. Web Inf. Syst. **5**(3), 1–22 (2009). https://doi.org/10.4018/jswis.2009081901
7. Blockeel, H., De Raedt, L.: Top-down induction of first-order logical decision trees. Artif. Intell. **101**(1–2), 285–297 (1998). https://doi.org/10.1016/S0004-3702(98)00034-4

8. Bloehdorn, S., Sure, Y.: Kernel methods for mining instance data in ontologies. In: Aberer, K., et al. (eds.) ASWC/ISWC -2007. LNCS, vol. 4825, pp. 58–71. Springer, Heidelberg (2007). https://doi.org/10.1007/978-3-540-76298-0_5

9. Bordes, A., Usunier, N., Garcia-Duran, A., Weston, J., Yakhnenko, O.: Translating embeddings for modeling multi-relational data. In: Burges, C.J.C., et al. (eds.) Proceedings of NIPS 2013, pp. 2787–2795. Curran Associates, Inc. (2013)

10. Cai, H., Zheng, V.W., Chang, K.: A comprehensive survey of graph embedding: Problems, techniques, and applications. IEEE Trans. Knowl. Data Eng. **30**(09), 1616–1637 (2018). https://doi.org/10.1109/TKDE.2018.2807452

11. Carbonneau, M., Cheplygina, V., Granger, E., Gagnon, G.: Multiple instance learning. Pattern Recogn. **77**, 329–353 (2018). https://doi.org/10.1016/j.patcog.2017.10.009

12. Chapelle, O., Schölkopf, B., Zien, A. (eds.): Semi-supervised Learning. The MIT Press, Cambridge (2006). https://doi.org/10.7551/mitpress/9780262033589.001.0001

13. Chen, J., Lécué, F., Pan, J., Horrocks, I., Chen, H.: Knowledge-based transfer learning explanation. In: Thielscher, M., et al. (eds.) Principles of Knowledge Representation and Reasoning: Proceedings of the Sixteenth International Conference, KR 2018, pp. 349–358. AAAI Press (2018)

14. Cochez, M., Ristoski, P., Ponzetto, S.P., Paulheim, H.: Global RDF vector space embeddings. In: d'Amato, C., et al. (eds.) ISWC 2017. LNCS, vol. 10587, pp. 190–207. Springer, Cham (2017). https://doi.org/10.1007/978-3-319-68288-4_12

15. d'Amato, C.: Logic and learning: Can we provide explanations in the current knowledge lake? In: Bonatti, P., et al. (eds.) Knowledge Graphs: New Directions for Knowledge Representation on the Semantic Web (Dagstuhl Seminar 18371), Dagstuhl Reports, vol. 8, pp. 37–38. Schloss Dagstuhl-Leibniz-Zentrum fuer Informatik (2019). https://doi.org/10.4230/DagRep.8.9.29

16. d'Amato, C., Fanizzi, N., Esposito, F.: Query answering and ontology population: an inductive approach. In: Bechhofer, S., Hauswirth, M., Hoffmann, J., Koubarakis, M. (eds.) ESWC 2008. LNCS, vol. 5021, pp. 288–302. Springer, Heidelberg (2008). https://doi.org/10.1007/978-3-540-68234-9_23

17. d'Amato, C., Fanizzi, N., Esposito, F.: Inductive learning for the semantic web: what does it buy? Semant. Web **1**(1–2), 53–59 (2010). https://doi.org/10.3233/SW-2010-0007

18. d'Amato, C., Quatraro, N.F., Fanizzi, N.: Injecting background knowledge into embedding models for predictive tasks on knowledge graphs. In: Verborgh, R., et al. (eds.) ESWC 2021. LNCS, vol. 12731, pp. 441–457. Springer, Cham (2021). https://doi.org/10.1007/978-3-030-77385-4_26

19. d'Amato, C., Tettamanzi, A.G.B., Minh, T.D.: Evolutionary discovery of multi-relational association rules from ontological knowledge bases. In: Blomqvist, E., Ciancarini, P., Poggi, F., Vitali, F. (eds.) EKAW 2016. LNCS (LNAI), vol. 10024, pp. 113–128. Springer, Cham (2016). https://doi.org/10.1007/978-3-319-49004-5_8

20. d'Amato, C.: Machine learning for the semantic web: lessons learnt and next research directions. Semant. Web **11**(1), 195–203 (2020). https://doi.org/10.3233/SW-200388

21. Deng, L., Yu, D. (eds.): Deep Learning: Methods and Applications. NOW Publishers, Delft (2014). https://doi.org/10.1561/2000000039

22. Doran, D., Schulz, S., Besold, T.: What does explainable AI really mean? A new conceptualization of perspectives. In: Besold, T.R., Kutz, O. (eds.) Proceedings of the First International Workshop on Comprehensibility and Explanation in AI and ML 2017 co-located with 16th International Conference of the Italian Association for Artificial Intelligence (AI*IA 2017), CEUR Work. Proc., vol. 2071. CEUR-WS.org (2017)

23. Fanizzi, N., d'Amato, C., Esposito, F.: Conceptual clustering and its application to concept drift and novelty detection. In: Bechhofer, S., Hauswirth, M., Hoffmann, J., Koubarakis, M.

(eds.) ESWC 2008. LNCS, vol. 5021, pp. 318–332. Springer, Heidelberg (2008). https://doi.org/10.1007/978-3-540-68234-9_25

24. Fanizzi, N., d'Amato, C., Esposito, F.: DL-FOIL concept learning in description logics. In: Železný, F., Lavrač, N. (eds.) ILP 2008. LNCS (LNAI), vol. 5194, pp. 107–121. Springer, Heidelberg (2008). https://doi.org/10.1007/978-3-540-85928-4_12

25. Fanizzi, N., d'Amato, C., Esposito, F.: Metric-based stochastic conceptual clustering for ontologies. Inf. Syst. **34**(8), 792–806 (2009). https://doi.org/10.1016/j.is.2009.03.008

26. Fanizzi, N., d'Amato, C., Esposito, F.: Induction of concepts in web ontologies through terminological decision trees. In: Balcázar, J.L., Bonchi, F., Gionis, A., Sebag, M. (eds.) ECML PKDD 2010. LNCS (LNAI), vol. 6321, pp. 442–457. Springer, Heidelberg (2010). https://doi.org/10.1007/978-3-642-15880-3_34

27. Fanizzi, N., Rizzo, G., d'Amato, C.: Boosting DL concept learners. In: Hitzler, P., et al. (eds.) ESWC 2019. LNCS, vol. 11503, pp. 68–83. Springer, Cham (2019). https://doi.org/10.1007/978-3-030-21348-0_5

28. Galárraga, L., Teflioudi, C., Hose, K., Suchanek, F.M.: AMIE: association rule mining under incomplete evidence in ontological knowledge bases. In: Schwabe, D., et al. (eds.) 22nd International World Wide Web Conference, WWW 2013, pp. 413–422. International World Wide Web Conferences Steering Committee/ACM (2013). https://doi.org/10.1145/2488388.2488425

29. Galárraga, L., Teflioudi, C., Hose, K., Suchanek, F.M.: Fast rule mining in ontological knowledge bases with AMIE+. The VLDB J. **24**(6), 707–730 (2015). https://doi.org/10.1007/s00778-015-0394-1

30. d'Avila Garcez, A., et al.: Neural-symbolic learning and reasoning: contributions and challenges. In: 2015 AAAI Spring Symposia. AAAI Press (2015). http://www.aaai.org/ocs/index.php/SSS/SSS15/paper/view/10281

31. Getoor, L., Taskar, B. (eds.): Introduction to Statistical Relational Learning. MIT Press, Cambridge (2007)

32. Guo, H., Herna, V.L.: Learning from imbalanced data sets with boosting and data generation: the databoost-im approach. SIGKDD Explor. **6**(1), 30–39 (2004). https://doi.org/10.1145/1007730.1007736

33. Guo, S., Wang, Q., Wang, L., Wang, B., Guo, L.: Jointly embedding knowledge graphs and logical rules. In: Proceedings of EMNLP 2016, pp. 192–202. ACL (2016). https://doi.org/10.18653/v1/D16-1019

34. Hitzler, P., Bianchi, F., Ebrahimi, M., Sarker, M.K.: Neural-symbolic integration and the semantic web. Semant. Web J. **11**(1), 3–11 (2020). https://doi.org/10.3233/SW-190368

35. Hoekstra, R.: The knowledge reengineering bottleneck. Semant. Web J. **1**, 111–115 (2010). https://doi.org/10.3233/SW-2010-0004

36. Hogan, A., et al.: Knowledge graphs. ACM Comput. Surv. **54**, 1–37 (2021). https://doi.org/10.1145/3447772

37. Horrocks, I., Patel-Schneider, P., Boley, H., Tabet, S., Grosof, B., Dean., M.: SWRL: a semantic web rule language combining owl and RuleML (2004). http://www.aaai.org/ocs/index.php/SSS/SSS15/paper/view/10281

38. Jayathilaka, M., Mu, T., Sattler, U.: Visual-semantic embedding model informed by structured knowledge. In: Rudolph, S., Marreiros, G. (eds.) Proceedings of STAIRS 2020. CEUR, vol. 2655. CEUR-WS.org (2020). http://ceur-ws.org/Vol-2655/paper23.pdf

39. Ji, S., Pan, S., Cambria, E., Marttinen, P., Yu, P.S.: A survey on knowledge graphs: representation, acquisition and applications (2020). arXiv:2002.00388

40. Józefowska, J., Lawrynowicz, A., Lukaszewski, T.: The role of semantics in mining frequent patterns from knowledge bases in description logics with rules. TPLP **10**(3), 251–289 (2010). https://doi.org/10.1017/S1471068410000098

41. Kazemi, S., Poole, D.: Simple embedding for link prediction in knowledge graphs. In: Bengio, S., et al. (eds.) Advances in Neural Information Processing Systems 31: Annual Conference on Neural Information Processing Systems 2018, NeurIPS 2018, pp. 4289–4300. ACM (2018)
42. Koller, D., Friedman, N. (eds.): Probabilistic Graphical Models: Principles and Techniques. MIT Press, Cambridge (2009)
43. Kramer, S., Lavrač, N., Flach, P.: Propositionalization approaches to relational data mining. In: Džeroski, S., Lavraž, N. (eds.) Relational Data Mining, pp. 262–291. LNCS, Springer (2001). https://doi.org/10.1007/978-3-662-04599-2_11
44. Labaf, M., Hitzler, P., Evans, A.: Propositional rule extraction from neural networks under background knowledge. In: Besold, T.R., et al. (eds.) Proceedings of the Twelfth International Workshop on Neural-Symbolic Learning and Reasoning, NeSy 2017. CEUR Workshop Proceedings, vol. 2003. CEUR-WS.org (2017)
45. Lehmann, J., Auer, S., Bühmann, L., Tramp, S.: Class expression learning for ontology engineering. J. Web Semant. 9(1), 71–81 (2011). https://doi.org/10.1016/j.websem.2011.01.001
46. Lehmann, J., Bühmann, L.: ORE - a tool for repairing and enriching knowledge bases. In: Patel-Schneider, P.F., et al. (eds.) ISWC 2010. LNCS, vol. 6497, pp. 177–193. Springer, Heidelberg (2010). https://doi.org/10.1007/978-3-642-17749-1_12
47. Lin, Y., Liu, Z., Sun, M., Liu, Y., Zhu, X.: Learning entity and relation embeddings for knowledge graph completion. In: AAAI 2015 Proceedings, pp. 2181–2187. AAAI Press (2015)
48. Liu, X., Wu, J., Zhou, Z.: Exploratory under-sampling for class-imbalance learning. In: Proceedings of the 6th IEEE International Conference on Data Mining (ICDM 2006), pp. 965–969. IEEE Computer Society (2006). https://doi.org/10.1109/ICDM.2006.68
49. Luger, G.F. (ed.): Artificial Intelligence: Structures and Strategies for Complex Problem Solving. Addison Wesley, Boston (2005)
50. Melo, A., Völker, J., Paulheim, H.: Type prediction in noisy RDF knowledge bases using hierarchical multilabel classification with graph and latent features. Int. J. Artif. Intell. Tools 26(2), 1–32 (2017). https://doi.org/10.1142/S0218213017600119
51. Minervini, P., d'Amato, C., Fanizzi, N.: Efficient energy-based embedding models for link prediction in knowledge graphs. J. Intell. Inf. Syst. 47(1), 91–109 (2016). https://doi.org/10.1007/s10844-016-0414-7
52. Minervini, P., Tresp, V., d'Amato, C., Fanizzi, N.: Adaptive knowledge propagation in web ontologies. TWEB 12(1), 2:1-2:28 (2018). https://doi.org/10.1145/3105961
53. Minervini, P., Costabello, L., Muñoz, E., Nováček, V., Vandenbussche, P.-Y.: Regularizing knowledge graph embeddings via equivalence and inversion axioms. In: Ceci, M., Hollmén, J., Todorovski, L., Vens, C., Džeroski, S. (eds.) ECML PKDD 2017. LNCS (LNAI), vol. 10534, pp. 668–683. Springer, Cham (2017). https://doi.org/10.1007/978-3-319-71249-9_40
54. Minervini, P., Demeester, T., Rocktäschel, T., Riedel, S.: Adversarial sets for regularising neural link predictors. In: Elidan, G., et al. (eds.) UAI 2017 Proceedings. AUAI Press (2017)
55. Nickel, M., Murphy, K., Tresp, V., Gabrilovich, E.: A review of relational machine learning for knowledge graphs. Proc. IEEE 104(1), 11–33 (2016). https://doi.org/10.1109/JPROC.2015.2483592
56. Nickel, M., Tresp, V., Kriegel, H.: A three-way model for collective learning on multi-relational data. In: Getoor, L., Scheffer, T. (eds.) Proceedings of the 28th International Conference on Machine Learning, ICML 2011, pp. 809–816. Omnipress (2011). https://icml.cc/2011/papers/438_icmlpaper.pdf
57. Paulheim, H.: Make embeddings semantic again! In: Proceedings of the ISWC 2018 P&D-Industry-BlueSky Tracks. CEUR Workshop Proceedings (2018)
58. Raedt, L.D. (ed.): Logical and Relational Learning: From ILP to MRDM (Cognitive Technologies). Springer-Verlag, Berlin (2008)

59. Rettinger, A., Lösch, U., Tresp, V., d'Amato, C., Fanizzi, N.: Mining the semantic web - statistical learning for next generation knowledge bases. Data Mining Knowl. Disc. **24**(3), 613–662 (2012). https://doi.org/10.1007/s10618-012-0253-2

60. Rettinger, A., Nickles, M., Tresp, V.: Statistical relational learning with formal ontologies. In: Buntine, W., Grobelnik, M., Mladenić, D., Shawe-Taylor, J. (eds.) ECML PKDD 2009. LNCS (LNAI), vol. 5782, pp. 286–301. Springer, Heidelberg (2009). https://doi.org/10.1007/978-3-642-04174-7_19

61. Ristoski, P., Paulheim, H.: RDF2Vec: RDF graph embeddings for data mining. In: Groth, P., et al. (eds.) ISWC 2016. LNCS, vol. 9981, pp. 498–514. Springer, Cham (2016). https://doi.org/10.1007/978-3-319-46523-4_30

62. Rizzo, G., d'Amato, C., Fanizzi, N., Esposito, F.: Terminological cluster trees for disjointness axiom discovery. In: Blomqvist, E., et al. (eds.) ESWC 2017. LNCS, vol. 10249, pp. 184–201. Springer, Cham (2017). https://doi.org/10.1007/978-3-319-58068-5_12

63. Rizzo, G., Fanizzi, N., d'Amato, C., Esposito, F.: Approximate classification with web ontologies through evidential terminological trees and forests. Int. J. Approx. Reason. **92**, 340–362 (2018). https://doi.org/10.1016/j.ijar.2017.10.019

64. Rizzo, G., Fanizzi, N., d'Amato, C., Esposito, F.: A framework for tackling myopia in concept learning on the web of data. In: Faron Zucker, C., Ghidini, C., Napoli, A., Toussaint, Y. (eds.) EKAW 2018. LNCS (LNAI), vol. 11313, pp. 338–354. Springer, Cham (2018). https://doi.org/10.1007/978-3-030-03667-6_22

65. Sarker, M., Xie, N., Doran, D., Raymer, M., Hitzler, P.: Explaining trained neural networks with semantic web technologies: First steps. In: Besold, T.R., et al. (eds.) Proceedings of the Twelfth International Workshop on Neural-Symbolic Learning and Reasoning, NeSy 2017. CEUR Workshop Proceedings, vol. 2003. CEUR-WS.org (2017)

66. Shadbolt, N., Berners-Lee, T., Hall, W.: The semantic web revisited. IEEE Intell. Syst. **21**(3), 96–101 (2006). https://doi.org/10.1109/MIS.2006.62

67. Siorpaes, K., Hepp, M.: OntoGame: towards overcoming the incentive bottleneck in ontology building. In: Meersman, R., Tari, Z., Herrero, P. (eds.) OTM 2007. LNCS, vol. 4806, pp. 1222–1232. Springer, Heidelberg (2007). https://doi.org/10.1007/978-3-540-76890-6_50

68. Spinosa, E., de Leon Ferreira de Carvalho, A.P., Gama, J.: Olindda: A cluster-based approach for detecting novelty and concept drift in data streams. In: Symposium of Applied Computing: Proceedings of the ACM International Conference, SAC 2007. vol. 1, pp. 448–452. ACM (2007)

69. Tiddi, I., d'Aquin, M., Motta, E.: Dedalo: looking for clusters explanations in a labyrinth of linked data. In: Presutti, V., et al. (eds.) ESWC 2014. LNCS, vol. 8465, pp. 333–348. Springer, Cham (2014). https://doi.org/10.1007/978-3-319-07443-6_23

70. Tran, A.C., Dietrich, J., Guesgen, H.W., Marsland, S.: An approach to parallel class expression learning. In: Bikakis, A., Giurca, A. (eds.) RuleML 2012. LNCS, vol. 7438, pp. 302–316. Springer, Heidelberg (2012). https://doi.org/10.1007/978-3-642-32689-9_25

71. Tran, A., Dietrich, J., Guesgen, H., Marsland, S.: Parallel symmetric class expression learning. J. Mach. Learn. Res. **18**(64), 1–34 (2017)

72. Völker, J., Niepert, M.: Statistical schema induction. In: ESWC 2011. LNCS, vol. 6643, pp. 124–138. Springer, Heidelberg (2011). https://doi.org/10.1007/978-3-642-21034-1_9

73. Völker, J., Fleischhacker, D., Stuckenschmidt, H.: Automatic acquisition of class disjointness. J. Web Semant. **35**(P2), 124–139 (2015). https://doi.org/10.1016/j.websem.2015.07.001

74. West, R., Gabrilovich, E., Murphy, K., Sun, S., Gupta, R., Lin, D.: Knowledge base completion via search-based question answering. In: Chung, C., et al. (eds.) 23rd International World Wide Web Conference, WWW 2014, pp. 515–526. ACM (2014). https://doi.org/10.1145/2566486.2568032

75. Widmer, G., Kubat, M.: Learning in the presence of concept drift and hidden contexts. Mach. Learn. **23**(1), 69–101 (1996)
76. Yang, B., Yih, W., He, X., Gao, J., Deng, L.: Embedding entities and relations for learning and inference in knowledge bases. In: Proceedings of ICLR 2015 (2015)
77. Zhou, Z., Zhang, M.: Multi-label learning. In: Sammut, C., Geoffrey, W. (eds.) Encyclopedia of Machine Learning and Data Mining, pp. 875–881. Springer, Berlin (2017). https://doi.org/ 10.1007/978-1-4899-7687-1_910

Temporal ASP: From Logical Foundations to Practical Use with `telingo`

Pedro Cabalar[✉]

University of Corunna, A Coruña, Spain
`cabalar@udc.es`

Abstract. This document contains some lecture notes for a seminar on Temporal Equilibrium Logic (TEL) and its application to Answer Set Programming (ASP) inside the 17th Reasoning Web Summer School (RW 2021). TEL is a temporal extension of ASP that introduces temporal modal operators as those from Linear-Time Temporal Logic. We present the basic definitions and intuitions for Equilibrium Logic and then extend these notions to the temporal case. We also introduce several examples using the temporal ASP tool `telingo`.

Keywords: Answer Set Programming · Linear Temporal Logic · Equilibrium Logic · Temporal Equilibrium Logic

1 Introduction

Answer Set Programming [4] (ASP) is nowadays one of the most successful paradigms for declarative problem solving and practical Knowledge Representation. Based on the answer set (or stable model) semantics [18] for logic programs, ASP constitutes a declarative formalism and a natural choice for solving static combinatorial problems up to NP complexity (or Σ_2^P in the disjunctive case), but has also been applied to problems that involve a dynamic component and a higher complexity, like planning, well-known to be PSPACE-complete [5]. The use of ASP for temporal scenarios has been frequent since its early application for reasoning about actions and change [19]. Commonly, dynamic scenarios in ASP deal with transition systems and discrete time: instants are represented as integer values for a time-point parameter, added to all dynamic predicates. This temporal parameter is bound to a finite interval, from 0 to a maximum time-step n (usually called the *horizon*). Problems involving temporal search, such as planning or temporal explanation, are solved by multiple calls to the ASP solver and gradually increasing the horizon length.

Although this methodology is simple and provides a high degree of flexibility, it lacks for a differentiated treatment of temporal expressions (the time parameter is just one more logical variable to be grounded) and prevents the reuse of

---

This research was partially supported by the Spanish Ministry of Economy and Competitivity (MINECO), grant TIN2017-84453-P.

© Springer Nature Switzerland AG 2022

M. Šimkus and I. Varzinczak (Eds.): Reasoning Web 2021, LNCS 13100, pp. 94–114, 2022.
https://doi.org/10.1007/978-3-030-95481-9_5

the large corpora of techniques and results well-known from the temporal logic literature. In an attempt to overcome these limitations, the approach called *Temporal Equilibrium Logic* [2,7] introduced a logical formalisation that combines ASP with modal temporal operators. This formalism constitutes an extension of *Equilibrium Logic* [22] which, in its turn, is a complete logical characterisation of (standard) ASP based on the intermediate logic of *Here-and-There* (HT) [21]. As a result, TEL is an expressive non-monotonic modal logic that shares the syntax of Linear-Time Temporal Logic (LTL) [23] but interprets temporal formulas under a non-monotonic semantics that properly extends stable models. This semantics is based on the idea of selecting some LTL temporal models of a theory Γ that satisfy some minimality condition, when examined under the weaker logic of temporal HT (THT). Thus, a temporal stable model of Γ is a kind of selected LTL model of Γ, and so, it has the form of a sequence of states, usually called a *trace*.

In the rest of this document we will first introduce some intuitions about Equilibrium Logic from a rule-based reasoning perspective and then shift to the definition of its temporal extension. After that, we will provide several examples and talk about their practical implementation in the ASP tool called `telingo`. Finally, we will close with some conclusions and open topics for future work or currently under study.

2   Rule-Based Reasoning and Equilibrium Logic

In this section, we partly reproduce the motivations included in [12] (see that paper for further detail). ASP is a rule-based paradigm sharing the same syntax as the logic programming language Prolog but with a different reading. Take a rule of the form:

$$\text{smoke :- fire.} \tag{1}$$

ASP uses a *bottom-up* reading of (1): "smoke is produced by fire". That is, whenever `fire` belongs to our current set of beliefs or certain facts, `smoke` must also be included in that set too. On the contrary, Prolog's *top-down* reading could be informally stated as "to obtain smoke, we need fire". That is, the rule describes a procedure to get `smoke` as a goal which consists in pursuing `fire` as a new goal. Regardless of the application direction, it seems clear that rules have a conditional form with a right-hand condition (*body*) and a left-hand consequent (*head*) that in our example (1) respectively correspond to `fire` and `smoke`. Thus, a straightforward logical formalisation would be understanding (1) as the implication *fire* → *smoke* in classical propositional logic. This guarantees, for instance, that if we add *fire* as a program fact, we will get *smoke* as a conclusion (by application of *modus ponens*). So, the "operational" aspect of rule (1) can be captured by classical implication. However, the *semantics* of a classical implication is not enough to cover the intuitive meaning of a program rule. If our program only contains (1) and we read it as a rule, it is clear that *fire* is not satisfied, since *no rule can yield* that atom, and so, *smoke* is not obtained either. However, implication *fire* → *smoke*, which amounts to the classically equivalent disjunction ¬*fire* ∨ *smoke*, has three classical models: ∅ (both atoms false), {*smoke*} and {*fire, smoke*}. Note that the two last models seem to consider situations in which *smoke* or *fire* could be arbitrarily *assumed as true*, even though the program provides *no way to prove them*. An important observation is that ∅ happens to be the smallest model (with respect to set inclusion). This model is interesting because, somehow, it reflects the principle of not adding arbitrary true atoms that we are not forced to believe, and it coincides with the expected meaning for a program just containing (1). The existence of a least classical model is, in fact, guaranteed for logic programs without negation (or disjunction), so-called *positive logic programs*, and so, it was adopted as the main semantics [14] for logic programming until the introduction of negation. However, when negation came into play, classical logic was revealed to be insufficient again, even under the premise of minimal models selection. Suppose we have a program Π_1 consisting of the rules:

$$\text{fill :- empty, not fire.} \tag{2}$$

$$\text{empty.} \tag{3}$$

where (2) means that we always *fill* our gas tank if it is *empty* and there is no evidence on *fire*, and (3) says that the tank is empty indeed. As before, *fire* cannot be proved (it is not head of any rule) and so, the condition of (2) is satisfied, producing *fill* as a result. The straightforward logical translation of (2) is *empty* ∧ ¬*fire* → *fill* that, in combination with fact (3), produces three models: $T_1 = \{empty, fill\}$, $T_2 = \{empty, fire\}$ and $T_3 = \{empty, fire, fill\}$. Unfortunately, there is no least classical model any more: both T_1 (the expected model) and T_2 are minimal with respect to set inclusion. After all, the previous implication is classically *equivalent* to *empty* → *fire* ∨ *fill* which does not capture the directional behaviour of rule (2). The undesired minimal model T_2 is assuming *fire* to be true, although there is no way to prove that fact in the program. So, apparently, classical logic is too weak for capturing the meaning of logic programs in the sense that it provides the expected model(s), but also accepts other models (like T_2 and T_3) in which some atoms are abitrarily assumed to be true but not "justified by the program".

Suppose we had a way to classify true atoms distinguishing between those just being an *assumption* (classical model T) and those being also *justified* or *proved* by program rules. In our intended models, the set of justified atoms should precisely coincide with the set of assumed ones in T. As an example, suppose our assumed atoms are $T_3 = \{empty, fire, smoke\}$. Any justification should include *empty* because of fact (3). However, rule (2) seems to be unapplicable, because we are currently assuming that *fire* is possibly true, *fire* ∈ T_3, and so 'not fire' is not acceptable – there is some (weak) evidence about fire. As a result, atom *fill* is not necessarily justified and we can only derive {*empty*}, which is strictly smaller than our initial assumption T_3. Something similar happens for assumption $T_2 = \{empty, fire\}$. If we take classical model $T_1 = \{empty, fill\}$ instead as an initial assumption, then the body of rule (2) becomes applicable, since no evidence on *fire* can be found, that is, $fire \notin T_1$. As a result, the justified atoms are now $\{empty, fill\} = T_1$ and the classical model T_1 becomes the unique intended (stable) model of the program.

The method we have just used with the example can be seen as an informal description of the original definition of the stable models semantics [18]. This definition consisted of classical logic reinforced with an *extra-logical* program transformation (for interpreting negation) and then using application of rules to obtain the actually derived or justified information. We show next how it is possible to provide an equivalent definition that exclusively resorts to logical concepts but using a different underlying formalism, weaker than classical logic.

Although, as we have seen, our interest is focused on rules, the semantics of Equilibrium Logic [22] can be defined on any arbitrary propositional formula. Covering arbitrary formulas is, in fact, simpler and more homogeneous than the original definition of stable models based on the reduct syntactic transformation. Equilibrium models are defined by a models selection criterion on top of the intermediate logic of Here-and-There (HT) [21], stronger than intuitionistic logic but weaker than classical logic. The latter can be seen as a three-valued logic where an atom can be false, assumed or proved, as we discussed before. Formally, an HT *interpretation* is a pair of sets of atoms $\langle H, T \rangle$ satisfying $H \subseteq T$ so that: any atom $p \in H$ is considered as *proved* or *justified*; an atom $p \in T$ is considered as *assumed*; and any atom $p \notin T$ is understood as *false*. As we can see, proved implies assumed, that is, the set of justified atoms H is always a subset of the assumed ones T. Intuitively, T acts as our "initial assumption" while the subset H contains those atoms from T currently considered as justified. Let At be the collection of all atomic formulas in our given language. Then $H \subseteq T \subseteq At$ and all atoms in $At \backslash T$ are considered false in this model. An HT interpretation $\langle H, T \rangle$ is said to be *total* when $H = T$ (that is, when all assumptions are justified).

As we did for atoms, formulas can also be considered to be false, assumed or proved. We will use a satisfaction relation $\langle H, T \rangle \models \varphi$ to represent that $\langle H, T \rangle$ makes formula φ to be proved or justified. Sometimes, however, we may happen that this relation does not hold $\langle H, T \rangle \not\models \varphi$ while in classical logic satisfaction $T \models \varphi$ using the assumptions in T is true. Then, we may say that the formula is just assumed. Finally, when φ is not even classically satisfied by T, $T \not\models \varphi$, we can guarantee that the formula is false. Formally, the fact that an interpretation $\langle H, T \rangle$ *satisfies* a formula φ (or makes it justified), written $\langle H, T \rangle \models \varphi$, is recursively defined as follows:

- $\langle H, T \rangle \not\models \bot$
- $\langle H, T \rangle \models p$ iff $p \in H$
- $\langle H, T \rangle \models \varphi \wedge \psi$ iff $\langle H, T \rangle \models \varphi$ and $\langle H, T \rangle \models \psi$
- $\langle H, T \rangle \models \varphi \vee \psi$ iff $\langle H, T \rangle \models \varphi$ or $\langle H, T \rangle \models \psi$
- $\langle H, T \rangle \models \varphi \rightarrow \psi$ iff both (i) $T \models \varphi$ implies $T \models \psi$ and (ii) $\langle H, T \rangle \models \varphi$ implies $\langle H, T \rangle \models \psi$

By abuse of notation, we use '$\models$' both for classical and for HT-satisfaction: the ambiguity is resolved by the form of the left interpretation (a single set T for classical and a pair $\langle H, T \rangle$ for HT). We say that an interpretation $\langle H, T \rangle$ is a model of a theory (set of formulas) Γ iff $\langle H, T \rangle \models \varphi$ for all $\varphi \in \Gamma$. We say that a propositional theory Γ *entails* some formula φ, written $\Gamma \models \varphi$, if any model of Γ is also a model of φ.

As we can see, everything is pretty standard excepting for the interpretation of implication, which imposes a stronger condition than in classical logic. In order to satisfy $\langle H, T \rangle \models \varphi \rightarrow \psi$, the standard condition would be (ii), that is, if the antecedent holds, then the consequent must hold too. In our case, the reading is closer to an application of *modus ponens* in an inference rule: if the antecedent is proved, then we can also prove the consequent. This condition, however, is further reinforced by (i) which informally means that our set of assumptions T classically satisfy the implication $\varphi \rightarrow \psi$ as well.

The following proposition tells us that satisfaction for total models amounts to classical satisfaction:

Proposition 1. *For any formula φ and set of atoms T, $\langle T, T \rangle \models \varphi$ iff $T \models \varphi$ in classical logic.*

We may read this result saying that classical models are a subset of HT models (they correspond to total HT models). This immediately means that any HT tautology is also a classical tautology. The opposite does not hold, namely, there are classical tautologies that are not HT tautologies. We will see later several examples.

Classical satisfaction for T allows us to keep the three-valued reading (*false*, *assumed* or *proved*) also for formulas in the following way. We say that $\langle H, T \rangle$ makes formula φ:

- *proved* when $\langle H, T \rangle \models \varphi$,
- *assumed* when $T \models \varphi$,
- *false* when $T \not\models \varphi$.

Interestingly, as happened with atoms, formulas also satisfy that anything proved must also be assumed. This is stated as the following property called *persistence*:

Proposition 2 (Persistence). *For any formula φ and any HT interpretation $\langle H, T \rangle$ we can show that, if φ is proved then it is also assumed, that is: $\langle H, T \rangle \models \varphi$ implies $T \models \varphi$.*

Notice that we did not provide satisfaction of negation $\neg \varphi$. This is because negation is not included above because it can be defined in terms of implication as the formula $\varphi \rightarrow \bot$, as happens in intuitionistic logic. Using that abbreviation and after some analysis, it can be proved that $\langle H, T \rangle \models \neg \varphi$ amounts to $T \not\models \varphi$, that is, $\neg \varphi$ is justified simply when φ is not assumed, that is, when it is false. Apart from negation, we also define the common Boolean operators $\top \overset{\text{def}}{=} \neg \bot$ and $\varphi \leftrightarrow \psi \overset{\text{def}}{=} (\varphi \rightarrow \psi) \wedge (\psi \rightarrow \varphi)$.

If we apply the persistence property to our example program Π_1, this means that any of its models $\langle H, T \rangle \models \Pi$ must satisfy $T \models \Pi$ as well. As we saw, we have only three possibilities for the latter, T_1, T_2 and T_3. On the other hand, the program fact (3) fixes $empty \in H$. Now, take assumption $T_1 = \{empty, fill\}$. The only model we get is $\langle T_1, T_1 \rangle$ because the other possible subset $H = \{empty\}$ of T_1 does not satisfy (2): $empty$ is justified, $fire$ is false, so we should get $fill$. Take T_2 instead. Apart from $\langle T_2, T_2 \rangle$, in this case we also get a model $\langle H, T_2 \rangle$ with $H = \{empty\}$. In such a case, $fire$ is only assumed true, but not proved. As a result, the rule is satisfied because its condition $\neg fire$ is false (we have some evidence on $fire$) and so $\langle \{empty\}, T_2 \rangle$ becomes a model. This is a clear evidence that our initial assumption adding $fire$ is not necessarily proved when we check the program rules. In the case of $T_3 = \{empty, fire, fill\}$ we have a similar situation. Interpretations with $H = \{empty\}$, $H = \{empty, fire\}$ or $H = \{empty, fill\}$ are also models. Note that in all of them, the only atom that is always proved is $empty$, pointing out again that $fire$ or $fill$ are not necessarily justified (cannot be proved using the program rules).

It must be understood that, at this point, the tag "justified" or "proved" just refers to a second kind of truth, stronger than "assumed." This tag will only acquire a real "provability" meaning once we introduce a models minimisation. For instance, for the formula $fire \rightarrow smoke$, we will have a model $H = T = \{smoke\}$ where $smoke$ is being considered justified. We still miss some minimisation criterion to consider justified or proved only those atoms and formulas that we are certain to be so. This minimisation selects some particular HT models and will follow the intuitive idea: given a fixed set of assumptions T, minimise proved atoms H. If we additionally require that anything assumed must be eventually proved, we get the following definition of *equilibrium models* first introduced by Pearce [22]:

Definition 1 (Equilibrium Model). *A total HT interpretation $\langle T, T \rangle$ is an equilibrium model of a theory Γ if $\langle T, T \rangle \models \Gamma$ and there is no $H \subset T$ such that $\langle H, T \rangle \models \Gamma$. When this happens, we also say that T is a stable model (or answer set) of Γ.*

From the logical point of view, it is now an easy task to define a stable model. The intuition is that we will be interested in cases where anything assumed true in set T eventually becomes necessarily proved, i.e., $H = T$ is the only possibility for assumption T.

Back to our example, the only stable model of Π_1 is the expected $T_1 = \{empty, fill\}$. This is because for the other two classical models $T_2 = \{empty, fire\}$ and $T_3 = \{empty, fill, fire\}$ we could see in the previous section that there were smaller sets H' that formed possible models of the program such as $\langle \{empty\}, T_2 \rangle$ or $\langle \{empty\}, T_3 \rangle$. In the case of T_1, however, the only obtained model is $\langle T_1, T_1 \rangle$ and no smaller $H \subset T_1$ can be used to form a model.

By selecting the equilibrium models, we obtain a non-monotonic entailment relation, that is, we may obtain conclusions that, after we add new information, can be retracted. For instance, in our example, if we are now said that $fire$ has been observed and we take program $\Pi_2 = \Pi_1 \cup \{fire\}$, the classical models become $T_4 = \{empty, fire\}$ and $T_5 = \{empty, fire, fill\}$. Clearly, T_4 will be a stable model, since there is no way to remove any of the two atoms $empty$, $fire$

that are now facts in program Π_2. However, T_5 is not in equilibrium, since we can form $H = \{empty, fire\}$ and $\langle H, T_5 \rangle \models \Pi_2$ because rule (2) is always satisfied, as its body is falsified since $T_5 \not\models \neg fire$. From the only stable model we obtain, we conclude that we cannot *fill* the tank any more, while this atom was among the previous conclusions when we had no information about *fire*.

As said before, some classical tautologies are not HT tautologies. An interesting example is the law of excluded middle $a \vee \neg a$. Intuitively, this formula means that a has to be proved or assumed and, accordingly, it has the HT countermodel $\langle \emptyset, \{a\} \rangle$. Therefore, if we include $a \vee \neg a$ for some atom in our theory, we force that assuming a is enough to consider it proved, so p will somehow behave "classically". Moreover, if we add the formula $a \vee \neg a$ for all atoms in the signature At, then all models are forced to be total $\langle T, T \rangle$ and Equilibrium Logic collapses into classical logic.

3   Temporal Equilibrium Logic

An important advantage of the definition of stable models based on Equilibrium Logic is that it provides a purely logical characterisation that has no syntactic limitations (it applies to arbitrary propositional formulas) and is easy to extend with the incorporation of new constructs or the definition of new combined logics. As happened with Equilibrium Logic, the definition of *(Linear-time) Temporal Equilibrium Logic* (TEL) is done in two steps. First, we define a monotonic temporal extension of HT, called *(Linear-time) Temporal Here-and-There* (THT) and, second, we select some models from THT that are said to be *in equilibrium*, obtaining in this way a non-monotonic entailment relation.

We reproduce next part of the contents from [1] and [10]. The original definition of TEL was thought as a direct non-monotonic extension of standard LTL, so that models had the form of infinite traces. However, this rules out computation by ASP technology and is unnatural for applications like planning, where plans amount to finite prefixes of one or more traces [13]. In a recent line of research [11], TEL was extended to cope with finite traces (which are closer to ASP computation). On the one hand, this amounts to a restriction of THT and TEL to finite traces. On the other hand, this is similar to the restriction of LTL to LTL_f advocated by [13]. Our new approach, dubbed TEL_f, has the following advantages. First, it is readily implementable via ASP technology. Second, it can be reduced to a normal form which is close to logic programs and much simpler than the one obtained for TEL. Finally, its temporal models are finite and offer a one-to-one correspondence to plans. Interestingly, TEL_f also sheds light on concepts and methodology used in incremental ASP solving when understanding incremental parameters as time points. Another distinctive feature of TEL_f is the inclusion of future as well as past temporal operators. When using the causal reading of program rules, it is generally more natural to draw upon the past in rule bodies and to refer to the future in rule heads. As well, past operators are much easier handled computationally than their future counterparts when it comes to incremental reasoning, since they refer to already computed knowledge.

In what follows, we present the general logics THT (monotonic) and TEL (non-monotonic) allowing traces of any length (possibly infinite), and will later on use the subindices ω or f to denote the particular cases where traces are always infinite or always finite, respectively. The syntax of THT (and TEL) is the same as for LTL with past operators. Given a (countable, possibly infinite) set At of propositional variables (called *alphabet*), *temporal formulas* φ are defined by the grammar:

$$\varphi ::= a \mid \bot \mid \varphi_1 \otimes \varphi_2 \mid \bullet\varphi \mid \varphi_1 \, \mathsf{S} \, \varphi_2 \mid \varphi_1 \, \mathsf{T} \, \varphi_2 \mid \mathsf{O}\varphi \mid \varphi_1 \, \mathbb{U} \, \varphi_2 \mid \varphi_1 \, \mathbb{R} \, \varphi_2 \mid \varphi_1 \, \mathbb{W} \, \varphi_2$$

where $a \in At$ is an atom and $\otimes$ is any binary Boolean connective $\otimes \in \{\rightarrow, \wedge, \vee\}$. The last six cases correspond to the temporal connectives whose names are listed below:

Past		Future	
$\bullet$ for *previous*		O for *next*	
S for *since*		$\mathbb{U}$ for *until*	
T for *trigger*		$\mathbb{R}$ for *release*	
		$\mathbb{W}$ for *while*	

We also define several common derived temporal operators:

$$\blacksquare\varphi \overset{\text{def}}{=} \bot \, \mathsf{T} \, \varphi \quad \textit{always before} \qquad \square\varphi \overset{\text{def}}{=} \bot \, \mathbb{R} \, \varphi \quad \textit{always afterward}$$

$$\blacklozenge\varphi \overset{\text{def}}{=} \top \, \mathsf{S} \, \varphi \quad \textit{eventually before} \qquad \lozenge\varphi \overset{\text{def}}{=} \top \, \mathbb{U} \, \varphi \quad \textit{eventually afterward}$$

$$\mathsf{I} \overset{\text{def}}{=} \neg\bullet\top \quad \textit{initial} \qquad \mathbb{F} \overset{\text{def}}{=} \neg\mathsf{O}\top \quad \textit{final}$$

$$\widehat{\bullet}\varphi \overset{\text{def}}{=} \bullet\varphi \vee \mathsf{I} \quad \textit{weak previous} \qquad \widehat{\mathsf{O}}\varphi \overset{\text{def}}{=} \mathsf{O}\varphi \vee \mathbb{F} \quad \textit{weak next}$$

A *(temporal) theory* is a (possibly infinite) set of temporal formulas. Note that we use solid operators to refer to the past, while future-time operators are denoted by outlined symbols.

As happens with HT with respect to classical logic, logics THT and LTL share the same syntax but, they have a different semantics, the former being a weaker logic than the latter. The semantics of LTL relies on the concept of a *trace*, a (possibly infinite) sequence of *states*, each of which is a set of atoms. For defining traces, we start by introducing some notation to deal with intervals of integer time points. Given $a \in \mathbb{N}$ and $b \in \mathbb{N} \cup \{\omega\}$, we let $[a..b]$ stand for the set $\{i \in \mathbb{N} \mid a \leq i \leq b\}$, $[a..b)$ for $\{i \in \mathbb{N} \mid a \leq i < b\}$ and $(a..b]$ for $\{i \in \mathbb{N} \mid a < i \leq b\}$. In LTL, a *trace* $\mathbf{T}$ of length λ over alphabet At is a sequence $\mathbf{T} = (T_i)_{i \in [0..\lambda)}$ of sets $T_i \subseteq At$. We sometimes use the notation $|\mathbf{T}| \overset{\text{def}}{=} \lambda$ to stand for the length of the trace. We say that $\mathbf{T}$ is *infinite* if $|\mathbf{T}| = \omega$ and *finite* if $|\mathbf{T}| \in \mathbb{N}$. To represent a given trace, we write a sequence of sets of atoms concatenated with '$\cdot$'. For instance, the finite trace $\{a\} \cdot \emptyset \cdot \{a\} \cdot \emptyset$ has length 4 and makes a true at even time points and false at odd ones. For infinite traces, we sometimes use ω-regular expressions like, for instance, in the infinite trace $(\{a\} \cdot \emptyset)^\omega$ where all even positions make a true and all odd positions make it false.

A state i is represented as a pair of sets of atoms $\langle H_i, T_i \rangle$ with $H_i \subseteq T_i \subseteq At$ where H_i (standing for "here") contains the proved atoms, whereas T_i (standing for "there") contains the assumed atoms. On the other hand, false atoms are just the ones not assumed, captured by $At \setminus T_i$. An HT-trace of length λ over alphabet At is a sequence of pairs $(\langle H_i, T_i \rangle)_{i \in [0..\lambda)}$ with $H_i \subseteq T_i$ for any $i \in [0..\lambda)$. For convenience, we usually represent the HT-trace as the pair $\langle \mathbf{H}, \mathbf{T} \rangle$ of traces $\mathbf{H} = (H_i)_{i \in [0..\lambda)}$ and $\mathbf{T} = (T_i)_{i \in [0..\lambda)}$. Given $\mathbf{M} = \langle \mathbf{H}, \mathbf{T} \rangle$, we also denote its length as $|\mathbf{M}| \overset{\text{def}}{=} |\mathbf{H}| = |\mathbf{T}| = \lambda$. Note that the two traces $\mathbf{H}, \mathbf{T}$ must satisfy a kind of order relation, since $H_i \subseteq T_i$ for each time point i. Formally, we define the ordering $\mathbf{H} \leq \mathbf{T}$ between two traces of the same length λ as $H_i \subseteq T_i$ for each $i \in [0..\lambda)$. Furthermore, we define $\mathbf{H} < \mathbf{T}$ as both $\mathbf{H} \leq \mathbf{T}$ and $\mathbf{H} \neq \mathbf{T}$. Thus, an HT-trace can also be defined as any pair $\langle \mathbf{H}, \mathbf{T} \rangle$ of traces such that $\mathbf{H} \leq \mathbf{T}$. The particular type of HT-traces satisfying $\mathbf{H} = \mathbf{T}$ are called *total*.

Given any HT-trace $\mathbf{M} = \langle \mathbf{H}, \mathbf{T} \rangle$, we define the THT satisfaction of formulas as follows.

Definition 2 (THT-satisfaction). *An HT-trace* $\mathbf{M} = \langle \mathbf{H}, \mathbf{T} \rangle$ *of length* λ *over alphabet* At *satisfies a temporal formula* φ *at time point* $k \in [0..\lambda)$, *written* $\mathbf{M}, k \models \varphi$, *if the following conditions hold:*

1. $\mathbf{M}, k \models \top$ *and* $\mathbf{M}, k \not\models \bot$
2. $\mathbf{M}, k \models a$ *if* $a \in H_k$ *for any atom* $a \in At$
3. $\mathbf{M}, k \models \varphi \wedge \psi$ *iff* $\mathbf{M}, k \models \varphi$ *and* $\mathbf{M}, k \models \psi$
4. $\mathbf{M}, k \models \varphi \vee \psi$ *iff* $\mathbf{M}, k \models \varphi$ *or* $\mathbf{M}, k \models \psi$
5. $\mathbf{M}, k \models \varphi \rightarrow \psi$ *iff* $\langle \mathbf{H}', \mathbf{T} \rangle, k \not\models \varphi$ *or* $\langle \mathbf{H}', \mathbf{T} \rangle, k \models \psi$, *for all* $\mathbf{H}' \in \{\mathbf{H}, \mathbf{T}\}$
6. $\mathbf{M}, k \models \bullet \varphi$ *iff* $k > 0$ *and* $\mathbf{M}, k{-}1 \models \varphi$
7. $\mathbf{M}, k \models \varphi \mathbf{S} \psi$ *iff for some* $j \in [0..k]$, *we have* $\mathbf{M}, j \models \psi$ *and* $\mathbf{M}, i \models \varphi$ *for all* $i \in (j..k]$
8. $\mathbf{M}, k \models \varphi \mathbf{T} \psi$ *iff for all* $j \in [0..k]$, *we have* $\mathbf{M}, j \models \psi$ *or* $\mathbf{M}, i \models \varphi$ *for some* $i \in (j..k]$
9. $\mathbf{M}, k \models \bigcirc \varphi$ *iff* $k + 1 < \lambda$ *and* $\mathbf{M}, k{+}1 \models \varphi$
10. $\mathbf{M}, k \models \varphi \mathbb{U} \psi$ *iff for some* $j \in [k..\lambda)$, *we have* $\mathbf{M}, j \models \psi$ *and* $\mathbf{M}, i \models \varphi$ *for all* $i \in [k..j)$
11. $\mathbf{M}, k \models \varphi \mathbb{R} \psi$ *iff for all* $j \in [k..\lambda)$, *we have* $\mathbf{M}, j \models \psi$ *or* $\mathbf{M}, i \models \varphi$ *for some* $i \in [k..j)$
12. $\mathbf{M}, k \models \varphi \mathbb{W} \psi$ *iff for all* $j \in [k..\lambda)$, *we have* $\langle \mathbf{H}', \mathbf{T} \rangle, j \models \varphi$ *or* $\langle \mathbf{H}', \mathbf{T} \rangle, i \not\models \psi$ *for some* $i \in [k..j)$ *and for all* $\mathbf{H}' \in \{\mathbf{H}, \mathbf{T}\}$. □

In general, these conditions inherit the interpretation of connectives from LTL (with past operators) with just a few differences. A first minor variation is that we allow traces of arbitrary length λ, including both infinite ($\lambda = \omega$) and finite ($\lambda \in \mathbb{N}$) traces. A second difference with respect to LTL is the new connective $\varphi \mathbb{W} \psi$ which is also a kind of temporally-iterated HT implication. Its intuitive reading is "keep doing φ while condition ψ holds." In LTL, $\varphi \mathbb{W} \psi$ would just amount to $\neg \psi \mathbb{R} \varphi$, but under HT semantics both formulas have a different meaning, as the latter may provide evidence for φ even though the condition ψ does not hold.

An HT-trace $\mathbf{M}$ is a *model* of a temporal theory Γ if $\mathbf{M}, 0 \models \varphi$ for all $\varphi \in \Gamma$. We write $THT(\Gamma, \lambda)$ to stand for the set of THT-models of length λ of a theory Γ, and define $THT(\Gamma) \stackrel{\text{def}}{=} THT(\Gamma, \omega) \cup \bigcup_{\lambda \in \mathbb{N}} THT(\Gamma, \lambda)$. That is, $THT(\Gamma)$ is the whole set of models of Γ of any length. For $\Gamma = \{\varphi\}$, we just write $THT(\varphi, \lambda)$ and $THT(\varphi)$. We can analogously define $LTL(\Gamma, \lambda)$, that is, the set of traces of length λ that satisfy theory Γ, and $LTL(\Gamma)$, that is, the LTL-models of Γ any length. We omit specifying LTL satisfaction since it coincides with THT when HT-traces are total.

Proposition 3 ([2,11]). *Let $\mathbf{T}$ be a trace of length λ, φ a temporal formula, and $k \in [0..\lambda)$ a time point.*
   Then, $\mathbf{T}, k \models \varphi$ in LTL iff $\langle \mathbf{T}, \mathbf{T} \rangle, k \models \varphi$. $\square$

In fact, total models can be forced by adding the following set of *excluded middle* axioms:

$$\square(a \vee \neg a) \qquad \text{for each atom } a \in At \text{ in the signature.} \qquad \text{(EM)}$$

Proposition 4 ([2,11]). *Let $\langle \mathbf{H}, \mathbf{T} \rangle$ be an HT-trace and (EM) the theory containing all excluded middle axioms for every atom $a \in At$. Then, $\langle \mathbf{H}, \mathbf{T} \rangle$ is a model of (EM) iff $\mathbf{H} = \mathbf{T}$.* $\square$

Satisfaction of derived operators can be easily deduced, as shown next.

Proposition 5 ([2,11]). *Let $\mathbf{M} = \langle \mathbf{H}, \mathbf{T} \rangle$ be an HT-trace of length λ over At. Given the respective definitions of derived operators, we get the following satisfaction conditions:*

12. $\mathbf{M}, k \models \mathsf{I}$ iff $k = 0$
13. $\mathbf{M}, k \models \hat{\bullet}\varphi$ iff $k = 0$ or $\mathbf{M}, k-1 \models \varphi$
14. $\mathbf{M}, k \models \blacklozenge\varphi$ iff $\mathbf{M}, i \models \varphi$ for some $i \in [0..k]$
15. $\mathbf{M}, k \models \blacksquare\varphi$ iff $\mathbf{M}, i \models \varphi$ for all $i \in [0..k]$
16. $\mathbf{M}, k \models \mathbb{F}$ iff $k + 1 = \lambda$
17. $\mathbf{M}, k \models \hat{\circ}\varphi$ iff $k + 1 = \lambda$ or $\mathbf{M}, k+1 \models \varphi$
18. $\mathbf{M}, k \models \Diamond\varphi$ iff $\mathbf{M}, i \models \varphi$ for some $i \in [k..\lambda)$
19. $\mathbf{M}, k \models \square\varphi$ iff $\mathbf{M}, i \models \varphi$ for all $i \in [k..\lambda)$

$\square$

Given a set of THT-models, we define the ones in equilibrium as follows.

Definition 3 (Temporal Equilibrium/Stable Model). *Let $\mathfrak{S}$ be some set of HT-traces. A total HT-trace $\langle \mathbf{T}, \mathbf{T} \rangle \in \mathfrak{S}$ is a temporal equilibrium model of $\mathfrak{S}$ iff there is no other $\mathbf{H} < \mathbf{T}$ such that $\langle \mathbf{H}, \mathbf{T} \rangle \in \mathfrak{S}$. The trace $\mathbf{T}$ is called a temporal stable model (TS-model) of $\mathfrak{S}$.* $\square$

We further talk about temporal equilibrium or temporal stable models of a theory Γ when $\mathfrak{S} = THT(\Gamma)$, respectively. Moreover, we write $TEL(\Gamma, \lambda)$ and $TEL(\Gamma)$ to stand for the temporal equilibrium models of $THT(\Gamma, \lambda)$ and

$THT(\Gamma)$ respectively. We write $TSM(\Gamma, \lambda)$ and $TSM(\Gamma)$ to stand for the corresponding sets of TS-models. One interesting observation is that, since temporal equilibrium models are total models $\langle \mathbf{T}, \mathbf{T} \rangle$, due to Proposition 3, we obtain $TSM(\Gamma, \lambda) \subseteq LTL(\Gamma, \lambda)$ that is, temporal stable models are a subset of LTL-models.

Temporal Equilibrium Logic (TEL) is the (non-monotonic) logic induced by temporal equilibrium models. We can also define the variants TEL_ω and TEL_f by applying the corresponding restriction to infinite and finite traces, respectively.

As an example of non-monotonicity, consider the formula

$$\square(\bullet \textit{loaded} \wedge \neg \textit{unloaded} \rightarrow \textit{loaded}) \tag{4}$$

that corresponds to the inertia for *loaded*, together with the fact *loaded*, describing the initial state for that fluent. Without entering into too much detail, this formula behaves as the logic program with the rules:

```
loaded(0).
loaded(T) :- loaded(T-1), not unloaded(T).
```

for any time point $T > 0$. As expected, for some fixed λ, we get a unique temporal stable model of the form $\{loaded\}^\lambda$. This entails that *loaded* is always true, viz. $\square loaded$, as there is no reason for *unloaded* to become true. Note that in the most general case of TEL, we actually get one stable model per each possible λ, including $\lambda = \omega$. Now, consider formula (4) along with $loaded \wedge \bigcirc\bigcirc unloaded$ which amounts to adding the fact unloaded(2). As expected, for each λ, the only temporal stable model now is $\mathbf{T} = \{loaded\} \cdot \{loaded\} \cdot \{unloaded\} \cdot \emptyset^\alpha$ where α can be $*$ or ω. Note that by making $\bigcirc\bigcirc unloaded$ true, we are also forcing $|\mathbf{T}| \geq 3$, that is, there are no temporal stable models (nor even THT-models) of length smaller than three. Thus, by adding the new information $\bigcirc\bigcirc unloaded$ some conclusions that could be derived before, such as $\square loaded$, are not derivable any more.

As an example emphasizing the behavior of finite traces, take the formula

$$\square(\neg a \rightarrow \bigcirc a) \tag{5}$$

which can be seen as a program rule "a(T + 1) :- not a(T)" for any natural number T. As expected, temporal stable models make a false in even states and true in odd ones. However, we cannot take finite traces making a false at the final state $\lambda - 1$, since the rule would force $\bigcirc a$ and this implies the existence of a successor state. As a result, the temporal stable models of this formula have the form $(\emptyset \cdot \{a\})^+$ for finite traces in TEL_f, or the infinite trace $(\emptyset \cdot \{a\})^\omega$ in TEL_ω.

Another interesting example is the temporal formula

$$\square(\neg \bigcirc a \rightarrow a) \wedge \square(\bigcirc a \rightarrow a).$$

The corresponding rules "a(T) :- not a(T + 1)" and "a(T) :- a(T + 1)" have no stable model [15] when grounded for all natural numbers T. This is because there is no way to build a finite proof for any a(T), as it depends on infinitely many next states to be evaluated. The same happens in TEL_ω, that is, we get no infinite temporal stable model. However in TEL_f, we can use the fact that $\circ a$ is always false in the last state. Then, $\square(\neg \circ a \to a)$ supports a in that state and therewith $\square(\circ a \to a)$ inductively supports a everywhere.

As an example of a temporal expression not so close to logic programming, consider the formula $\square\diamond a$, which is normally used in LTL_ω to assert that a occurs infinitely often. As discussed by [13], if we assume finite traces, then the formula collapses to $\square(\mathbb{F} \to a)$ in LTL_f, that is, a is true at the final state (and either true or false everywhere else). The same behavior is obtained in THT_ω and THT_f, respectively. However, if we move to TEL, a truth minimization is additionally required. As a result, in TEL_f, we obtain a unique temporal stable model for each fixed $\lambda \in \mathbb{N}$, in which a is true at the last state, and false everywhere else. Unlike this, TEL_ω yields no temporal stable model at all. This is because for any $\mathbf{T}$ with an infinite number of a's we can always take some $\mathbf{H}$ from which we remove a at some state, and still have an infinite number of a's in $\mathbf{H}$. Thus, for any total THT_ω-model $\langle \mathbf{T}, \mathbf{T} \rangle$ of $\square\diamond a$ there always exists some model $\langle \mathbf{H}, \mathbf{T} \rangle$ with strictly smaller $\mathbf{H} < \mathbf{T}$. Note that we can still specify infinite traces with an infinite number of occurrences of a, but at the price of *removing the truth minimization* for that atom. This can be done, for instance, by adding the excluded middle axiom (EM) for atom a. In this way, infinite traces satisfying $\square\diamond a \wedge \square(a \vee \neg a)$ are those that contain an infinite number of a's. In fact, if we add the excluded middle axiom for all atoms, TEL collapses into LTL, as stated below.

Proposition 6. *Let Γ be a temporal theory over At and (EM) be the set of all excluded middle axioms for all atoms in At.*
Then, $TSM(\ \Gamma \cup (EM)\ ) = LTL(\Gamma)$. □

4   Computing Temporal Stable Models

Let us consider a more meaningful example, taking the Yale Shooting scenario [20] where we must shoot a loaded gun to kill a turkey. A possible encoding in TEL could be:

$$\square(loaded \wedge \circ shoot \to \circ dead) \tag{6}$$

$$\square(loaded \wedge \circ shoot \to \circ unloaded) \tag{7}$$

$$\square(load \to loaded) \tag{8}$$

$$\square(dead \to \circ dead) \tag{9}$$

$$\square(loaded \wedge \neg \circ unloaded \to \circ loaded) \tag{10}$$

$$\square(unloaded \wedge \neg \circ loaded \to \circ unloaded) \tag{11}$$

In this way, under TEL semantics, implication $\alpha \to \beta$ has a similar behaviour to a directional inference rule, normally reversed as $\beta \leftarrow \alpha$ or $\beta :- \alpha$ in logic programming notation. The last two rules, (10)–(11), encode the *inertia law* for fluents *loaded* and *unloaded*, respectively. Note the use of ¬ in these two rules: it actually corresponds to *default negation*, that is, ¬α is read as "there is no evidence about α." For instance, (10) is read as "if the gun was loaded and we cannot prove that it will become unloaded then it stays loaded".

Computation of temporal stable models is a complex task. THT-satisfiability has been classified [8] as PSPACE-complete, that is, the same complexity as LTL-satisfiability, whereas TEL-satisfiability rises to EXPSPACE-completeness, as proved in [3]. In this way, we face a similar situation as in the non-temporal case where HT-satisfiability is NP-complete like SAT, whereas existence of equilibrium model (for arbitrary theories) is Σ_2^P-complete (like disjunctive ASP). There exist a pair of tools, STeLP [6] and ABSTEM [9], that allow computing (infinite) temporal stable models (represented as Büchi automata). These tools can be used to check verification properties that are usual in LTL, like the typical safety, liveness and fairness conditions, but in the context of temporal ASP. Moreover, they can also be applied for planning problems that involve an indeterminate or even infinite number of steps, such as the non-existence of a plan. In most practical problems, however, we are normally interested in finite traces. For that purpose, TEL$_f$ is implemented in the telingo system, extending the ASP system clingo to compute the temporal stable models of (non-ground) temporal logic programs. To this end, it extends the full-fledged input language of clingo with temporal operators and computes temporal models incrementally by multi-shot solving using a modular translation into ASP. telingo is freely available at github[1]. For instance, under telingo syntax, our theory (6)–(11) would be represented[2] as

```
#program dynamic.
dead :- shoot, 'loaded.
unloaded :- shoot, 'unloaded.
loaded :- load.
dead :- 'dead.
loaded :- 'loaded, not unloaded.
unloaded :- 'unloaded, not loaded.
```

The telingo input language actually allows the introduction of arbitrary LTL formulas in constraints or past formulas in the rule bodies (conditions). The syntax extends the full-fledged modeling language of clingo by the future and past temporal operators listed in the first and fourth row of Table 1. To support incremental ASP solving, telingo accepts a fragment of TEL$_f$ called *past-future rules* (see [11] for more details). A temporal formula is a past-future rule if it has form *Hd* ← *Bd* where *Bd* and *Hd* are just temporal formulas with the following restrictions: *Bd* and *Hd* contain no implications (other than

---

[1] https://github.com/potassco/telingo.

[2] The left upper commas are read as *previously* and correspond to the past operator dual of *next* 'O'. The □ operator is implicit in all dynamic rules.

Table 1. Past and future temporal operators in `telingo` and TEL_f

`&initial`	I	*initial*	`&final`	$\mathbb{F}$	*final*
`'p`	$\bullet p$	*previous*	`p'`	$\mathrm{O} p$	*next*
`<`	$\bullet$	*previous*	`>`	O	*next*
`<?`	S	*since*	`>?`	$\mathbb{U}$	*until*
`<*`	T	*trigger*	`>*`	$\mathbb{R}$	*release*
`<?`	$\blacklozenge$	*eventually before*	`>?`	$\Diamond$	*eventually afterward*
`<*`	$\blacksquare$	*always before*	`>*`	$\Box$	*always afterward*
`<:`	$\widehat{\bullet}$	*weak previous*	`>:`	$\widehat{\mathrm{O}}$	*weak next*

negations[3]), Bd contains no future operators, and Hd contains no past operators. An example of a past-future rule is, for instance, the formula

$$\Box(shoot \wedge \bullet\blacklozenge shoot \wedge \blacksquare unloaded \rightarrow \Diamond fail) \qquad (12)$$

expressing the sentence: *"If we shoot twice with a gun that was never loaded, it will eventually fail."* The past-future fragment is not only quite expressive but also rather natural when using the causal reading of program rules by drawing upon the past in rule bodies and referring to the future in rule heads. Considering that, past-future rules also serve as the design guideline for `telingo`'s input language.

To this end, `telingo` allows for enclosing a nested temporal formula φ in an expression of the form `&tel{`φ`}`. Formulas like φ are formed via the temporal operators in Line 3 to 8 in Table 1 along with the Boolean operators `&`, `|`, `~` for conjunction, disjunction, and negation, respectively (thus avoiding nested implications). The underlying idea is to use the *smaller* symbol `<` as the basis of all past operators, and to combine it with a *question mark* `?` or a *Kleene star* `*` depending on whether the semantics of the respective operator relies on an existential or universal quantification over states. This is nicely exemplified by the always and eventually operators, represented by `<*` and `<?`. In fact, the symbols `<*` and `<?` are overloaded due to their usage as binary and unary operators. For a simple example, consider the formula $\bullet p \vee \blacklozenge r$ represented as '`&tel{< p | <? p}`'. Similarly, future operators are built with the *greater* symbol '`>`' as their basis. More generally, temporal expressions of the form `&tel{`φ`}` are treated like atoms in `telingo`'s input language (and constitute theory atoms in `clingo` [17]); they are compiled away by `telingo`'s preprocessing that ultimately yields present-centered logic programs. In order to keep this translation simple, the current version of `telingo`, viz 2.1.1, restricts their occurrence in temporal rules $Hd \leftarrow Bd$ to being positive in Hd and preceded by one or two negations in their body Bd.[4] No restriction is imposed on their occurrences in integrity constraints.

---

[3] Recall that $\neg\varphi \overset{\text{def}}{=} \varphi \rightarrow \bot$ in the logic of here-and-there and thus in TEL_f, too.

[4] The extension to arbitrary occurrences is no hurdle and foreseen in future versions of `telingo`.

For example, the integrity constraint '$shoot \wedge \blacksquare unloaded \wedge \bullet\blacklozenge shoot \rightarrow \bot$' is expressible in several alternative ways.

```
:- &tel { shoot & <* unloaded & < <? shoot }.
:- shoot,  &tel { <* unloaded & < <? shoot }.
:- shoot,  &tel { <* unloaded }, &tel { < <? shoot }.
```

Alternatively, present-centered logic programs can be written directly by using the alternative notation for the common one-step operators $\bullet$ and $\circ$. Here, a quote is used either at the beginning or the end of a predicate symbol to indicate that the literal at hand must be true in the previous or next state in the trace, respectively. For instance, $\bullet p(7)$ is represented by `'p(7)`, while $\circ q(X)$ is `q'(X)`. For convenience, `telingo` 2.1.1 allows for using $\circ$ in singleton rule heads;[5] as above, this is compiled away during preprocessing.

The distinction between different types of temporal rules is done in `telingo` via `clingo`'s #program directives [16], which allow us to partition programs into subprograms. More precisely, each rule in `telingo`'s input language is associated with a temporal rule r of form $Hd \leftarrow Bd$ and interpreted as r, $\widehat{\circ}\Box r$, or $\Box(\mathbb{F} \rightarrow r)$ depending on whether it occurs in the scope of a program declaration headed by initial, dynamic, or final, respectively. Additionally, `telingo` offers always for gathering rules preceded by $\Box$ (thus dropping $\widehat{\circ}$ from dynamic rules). A rule outside any such declaration is regarded to be in the scope of initial.

For illustration, we give in Listing 1.1 an exemplary `telingo` encoding of the *Fox, Goose and Beans Puzzle* available at https://github.com/potassco/telingo/tree/master/examples/river-crossing.

Once upon a time a farmer went to a market and purchased a fox, a goose, and a bag of beans. On his way home, the farmer came to the bank of a river and rented a boat. But crossing the river by boat, the farmer could carry only himself and a single one of his purchases: the fox, the goose, or the bag of beans. If left unattended together, the fox would eat the goose, or the goose would eat the beans. The farmer's challenge was to carry himself and his purchases to the far bank of the river, leaving each purchase intact. How did he do it?

(https://en.wikipedia.org/wiki/Fox,_goose_and_bag_of_beans_puzzle)

In Listing 1.1, lines 3–5 and 9–10 provide facts holding in all and the initial states, respectively; this is indicated by the program directives headed by always and initial. The dynamic rules in lines 14–22 describe the transition function. The farmer moves at each time step (Line 14), and may take an item or not (Line 15). Line 17 describes the effect of action move/1, Line 18 its precondition, and Line 20 the law of inertia. The second part of the always rules give state constraints in Line 24 and 25. The final rule in Line 29 gives the goal condition.

All in all, we obtain two shortest plans consisting of eight states in about 20 ms. Restricted to the move predicate, `telingo` reports the following solutions:

---

[5] As above, the extension to disjunctions is no principal hurdle and foreseen in future versions of `telingo`; currently they must be expressed by using &tel.

Listing 1.1. telingo encoding for the Fox, Goose and Beans Puzzle

```
#program always.

item(fox;beans;goose).
route(river_bank,far_bank). route(far_bank,river_bank).
eats(fox,goose). eats(goose,beans).

#program initial.

at(farmer,river_bank).
at(X,river_bank) :- item(X).

#program dynamic.

move(farmer).
0 { move(X) : item(X) } 1.

at(X,B) :- 'at(X,A), move(X), route(A,B).
:- move(X), item(X), 'at(farmer,A), not 'at(X,A).

at(X,A) :- 'at(X,A), not move(X).

#program always.

:- at(X,A), at(X,B), A<B.
:- eats(X,Y), at(X,A), at(Y,A), not at(farmer,A).

#program final.

:- at(X,river_bank).

#show move/1.
#show at/2.
```

Time	Solution 1		Solution 2	
1				
2	move(farmer)	move(goose)	move(farmer)	move(goose)
3	move(farmer)		move(farmer)	
4	move(beans)	move(farmer)	move(farmer)	move(fox)
5	move(farmer)	move(goose)	move(farmer)	move(goose)
6	move(farmer)	move(fox)	move(beans)	move(farmer)
7	move(farmer)		move(farmer)	
8	move(farmer)	move(goose)	move(farmer)	move(goose)

We have chosen this example since it was also used by [6] to illustrate the working of STeLP, a tool for temporal answer set programming with TEL_ω. We note that STeLP and telingo differ syntactically in describing transitions by using next or previous operators, respectively. Since telingo extends clingo's input language, it offers a richer input language, as witnessed by the cardinality constraints in Line 15 in Listing 1.1. Finally, STeLP uses a model checker and outputs an automaton capturing all infinite traces while telingo returns finite traces corresponding to plans.

As a second example, consider the following problem[6] proposed by Professor J. Moore from the University of Texas at Austin and submitted to the Texas Action Group (TAG) discussion group.

> *Consider two processes, A and B, each of which is reading and writing a shared variable C. Each process is in an infinite loop, repeatedly executing: C = C + C; By this we mean "read the value of C, read the value of C again, add the two results and store the sum in C." The two reads store the values in "local registers" of the process. Reads and writes are atomic but there is no synchronization between the two processes. The initial value of C is 1. Problem: Given an arbitrary positive integer n is there an execution that assigns C the value n?*

A first possible encoding of this problem in telingo could look simply be the one shown in Listing 1.2. In this encoding, action fetch(P) reflects the fact that the CPU has non-deterministically decided to execute the next instruction from process P, encoding the process interleaving in that way. This non-deterministic choice is encoded in lines 23–26. The most significant feature of this encoding is that we keep an auxiliary fluent i(P) to stand for the current *instruction pointer* of each process P. An interesting observation is that, once we allow temporal expressions in the rule bodies and constraints, we can sometimes replace auxiliary fluents in favour of temporal queries about the past execution. In our example, this leads to a second encoding shown in Listing 1.3. In this case, we have removed the fluent capturing the instruction pointer and replaced it by the constraints in lines 33–35. Lines 33–34 mean that if we fetch instruction I for a process P already fetched in the past, then the last instruction fetched for P (whenever this is located in the past) must be I+2 modulo 3 (that is, the previous one in the cyclic order of instructions 0, 1, 2, 0, ...). Line 35 forces that the first instruction to be executed by any process P is 0: we cannot fetch instructions 1 or 2 if process P has not been already fetched.

Listing 1.2. telingo basic encoding for Moore's problem

```
1   #const n=19.
2
3   #program initial.
4
5   process(a;b).
6   local(0;1).
7   instruction(0..2).
```

---

[6] https://www.cs.utexas.edu/users/vl/tag/jmoore_discussion.

```
 8
 9    % Each process P executes:
10    % 0    assign c to r(P)
11    % 1    add c to r(P)
12    % 2    assing r(P) to C
13    %
14    % i(P)=I points to the next instruction I of process P to execute
15
16    holds(i(P),0) :- process(P).
17    holds(c,1).
18    holds(r(P),0) :- process(P).
19
20    #program dynamic.
21
22    %1 {fetch(P): _process(P)} 1.
23    {fetch(P,I):_instruction(I)} 1 :- _process(P).
24    fetch(P) :- fetch(P,I).
25    :- fetch(P,I), not _local(I), fetch(Q), P!=Q.
26    :- #count{P:fetch(P)}=0.
27
28    change(i(P),(I+1)\3) :- fetch(P), 'holds(i(P),I).
29    change(r(P),C  )     :- fetch(P), 'holds(i(P),0), 'holds(c,C).
30    change(r(P),R+C)     :- fetch(P), 'holds(i(P),1), 'holds(c,C),
31                            'holds(r(P),R), R+C <= n.
32    change(r(P),n+1)     :- fetch(P), 'holds(i(P),1), 'holds(c,C),
33                            'holds(r(P),R), R+C >  n.
34    change(c    ,R  )    :- fetch(P), 'holds(i(P),2), 'holds(r(P),R).
35
36    holds(F,V) :- change(F,V).
37    holds(F,V) :- 'holds(F,V), not change(F,_).
38
39    #program final.
40    :- not _testing, not holds(c,n).
41
42    #show fetch/1.
43    #show holds/2.
```

Listing 1.3. `telingo` second encoding for Moore's problem

```
 1    #const n=23.
 2
 3    #program initial.
 4
 5    process(a;b).
 6    local(0;1).
 7    instruction(0..2).
 8
 9    % Each process P executes:
10    % 0    assign c to r(P)
11    % 1    add c to r(P)
12    % 2    assing r(P) to C
13
14    holds(c,1).
15    holds(r(P),0) :- process(P).
16
17    #program dynamic.
18
19    {fetch(P,I): _instruction(I)} 1 :- _process(P).
20    fetch(P) :- fetch(P,I).
21    :- fetch(P,I), not _local(I), fetch(Q), P!=Q.
22    :- #count{P:fetch(P)}=0.
```

```
23
24
25    change(r(P),C  ) :- fetch(P,0), 'holds(c,C).
26    change(r(P),R+C) :- fetch(P,1), 'holds(c,C), 'holds(r(P),R), R+C<=n.
27    change(r(P),n+1) :- fetch(P,1), 'holds(c,C), 'holds(r(P),R), R+C>n.
28    change(c   ,R  ) :- fetch(P,2), 'holds(r(P),R).
29
30    holds(F,V) :- change(F,V).
31    holds(F,V) :- 'holds(F,V), not change(F,_).
32
33    :- fetch(P,I), &tel{< <? fetch(P)},
34       not &tel { < (~fetch(P) <? fetch(P,I')) : I'=(I+2)\3 }.
35    :- fetch(P,I), not &tel{< <? fetch(P)}, I!=0.
36
37    #program final.
38    :- not _testing, not holds(c,n).
39
40    #show fetch/2.
41    #show holds/2.
```

5   Conclusions

These recent results open several interesting topics for future study. First, it would be interesting to adapt existing model checking techniques (based on automata construction) for temporal logics to solve the problem of existence of temporal stable models. This was done for infinite traces in [6,8], but no similar method has been implemented for finite traces on TEL_f. The importance of having an efficient implementation of such a method is that it would allow deciding non-existence of a plan in a given planning problem, something not possible by current incremental solving techniques. Another interesting topic is the optimization of grounding in temporal ASP specifications as those handled by telingo. The current grounding of telingo is inherited from incremental solving in clingo and does not exploit the semantics of temporal expressions that are available now in the input language. Finally, we envisage to extend the telingo system with features of DEL (an extension to cope with dynamic logic operators) in order to obtain a powerful system for representing and reasoning about dynamic domains, not only providing an effective implementation of TEL and DEL but, furthermore, a platform for action and control languages.

Acknowledgements. This document contains partial reproductions of joint work with my coauthors Felicidad Aguado, Martín Diéguez, Jorge Fandinno, Roland Kaminski, Philip Morkisch, David Pearce, Gilberto Pérez, Torsten Schaub, Anna Schuhmann, Agustín Valverde and Concepción Vidal, all of them involved in the development of Equilibrium Logic, particularly in its temporal extension and its application to practical ASP solving. I wish to thank all of them for the fruitful collaboration and friendship along these years and extend this recognition to other colleagues that actively participate in the discussions or development of linear-temporal ASP like François Laferriere, Susana Hahn, Etienne Tignon and Javier Romero from the Potassco group at Potsdam and Philip Balbiani, Andreas Herzig and Luis Fariñas del Cerro from the IRIT at Toulouse, among others.

References

1. Aguado, F., et al.: Linear-time temporal answer set programming. Theory Pract. Log. Program. (2021, submitted)
2. Aguado, F., Cabalar, P., Diéguez, M., Pérez, G., Vidal, C.: Temporal equilibrium logic: a survey. J. Appl. Non-Classical Log. **23**(1–2), 2–24 (2013)
3. Bozzelli, L., Pearce, D.: On the complexity of temporal equilibrium logic. In: Proceedings of the 30th Annual ACM/IEEE Symposium of Logic in Computer Science (LICS 2015), Kyoto, Japan (2015, to appear)
4. Brewka, G., Eiter, T., Truszczyński, M.: Answer set programming at a glance. Commun. ACM **54**(12), 92–103 (2011)
5. Bylander, T.: The computational complexity of propositional strips planning. Artif. Intell. **69**(1), 165–204 (1994)
6. Cabalar, P., Diéguez, M.: STELP – a tool for temporal answer set programming. In: Delgrande, J.P., Faber, W. (eds.) LPNMR 2011. LNCS (LNAI), vol. 6645, pp. 370–375. Springer, Heidelberg (2011). https://doi.org/10.1007/978-3-642-20895-9_43
7. Cabalar, P., Pérez Vega, G.: Temporal equilibrium logic: a first approach. In: Moreno Díaz, R., Pichler, F., Quesada Arencibia, A. (eds.) EUROCAST 2007. LNCS, vol. 4739, pp. 241–248. Springer, Heidelberg (2007). https://doi.org/10.1007/978-3-540-75867-9_31
8. Cabalar, P., Demri, S.: Automata-based computation of temporal equilibrium models. In: Vidal, G. (ed.) LOPSTR 2011. LNCS, vol. 7225, pp. 57–72. Springer, Heidelberg (2012). https://doi.org/10.1007/978-3-642-32211-2_5
9. Cabalar, P., Diéguez, M.: Strong equivalence of non-monotonic temporal theories. In: Proceedings of the 14th International Conference on Principles of Knowledge Representation and Reasoning (KR 2014), Vienna, Austria (2014)
10. Cabalar, P., Kaminski, R., Morkisch, P., Schaub, T.: telingo = ASP + time. In: Balduccini, M., Lierler, Y., Woltran, S. (eds.) LPNMR 2019. LNCS, vol. 11481, pp. 256–269. Springer, Cham (2019). https://doi.org/10.1007/978-3-030-20528-7_19
11. Cabalar, P., Kaminski, R., Schaub, T., Schuhmann, A.: Temporal answer set programming on finite traces. Theory Pract. Log. Program. **18**(3–4), 406–420 (2018)
12. Cabalar, P., Pearce, D., Valverde, A.: Answer set programming from a logical point of view. Künstl. Intell. **32**(2–3), 109–118 (2018)
13. De Giacomo, G., Vardi, M.: Linear temporal logic and linear dynamic logic on finite traces. In: Rossi, F. (ed.) Proceedings of the Twenty-third International Joint Conference on Artificial Intelligence (IJCAI 2013), pp. 854–860. IJCAI/AAAI Press (2013)
14. van Emden, M.H., Kowalski, R.A.: The semantics of predicate logic as a programming language. J. ACM **23**, 733–742 (1976)
15. Fages, F.: Consistency of Clark's completion and the existence of stable models. J. Methods Log. Comput. Sci. **1**, 51–60 (1994)
16. Gebser, M., Kaminski, R., Kaufmann, B., Schaub, T.: Multi-shot ASP solving with clingo. Theory Pract. Log. Program. **19**(1), 27–82 (2019). http://arxiv.org/abs/1705.09811

17. Gebser, M., Kaminski, R., Kaufmann, B., Ostrowski, M., Schaub, T., Wanko, P.: Theory solving made easy with Clingo 5. In: Carro, M., King, A., Saeedloei, N., Vos, M.D. (eds.) Technical Communications of the 32nd International Conference on Logic Programming (ICLP 2016). OpenAccess Series in Informatics (OASIcs), vol. 52, pp. 2:1–2:15. Schloss Dagstuhl-Leibniz-Zentrum fuer Informatik, Dagstuhl (2016). https://doi.org/10.4230/OASIcs.ICLP.2016.2. http://drops.dagstuhl.de/opus/volltexte/2016/6733

18. Gelfond, M., Lifschitz, V.: The stable model semantics for logic programming. In: Proceedings of the 5th International Conference on Logic Programming (ICLP 1988), Seattle, Washington, pp. 1070–1080 (1988)

19. Gelfond, M., Lifschitz, V.: Representing action and change by logic programs. J. Log. Program. **17**(2/3&4), 301–321 (1993)

20. Hanks, S., McDermott, D.V.: Nonmonotonic logic and temporal projection. Artif. Intell. **33**(3), 379–412 (1987)

21. Heyting, A.: Die formalen Regeln der intuitionistischen Logik. Sitzungsberichte der Preussischen Akademie der Wissenschaften. Physikalisch-mathematische Klasse (1930)

22. Pearce, D.: A new logical characterisation of stable models and answer sets. In: Dix, J., Pereira, L.M., Przymusinski, T.C. (eds.) NMELP 1996. LNCS, vol. 1216, pp. 57–70. Springer, Heidelberg (1997). https://doi.org/10.1007/BFb0023801

23. Pnueli, A.: The temporal logic of programs. In: 18th Annual Symposium on Foundations of Computer Science, pp. 46–57. IEEE Computer Society Press (1977)

A Review of SHACL: From Data Validation to Schema Reasoning for RDF Graphs

Paolo Pareti[1]([✉])[iD] and George Konstantinidis[2][iD]

[1] University of Winchester, Winchester, UK
paolo.pareti@winchester.ac.uk
[2] University of Southampton, Southampton, UK
g.konstantinidis@soton.ac.uk

Abstract. We present an introduction and a review of *Shapes Constraint Language* (SHACL), the W3C recommendation language for validating RDF data. A SHACL document describes a set of constraints on RDF nodes, and a graph is valid with respect to the document if its nodes satisfy these constraints. We revisit the basic concepts of the language, its constructs and components and their interaction. We review the different formal frameworks used to study this language and the different semantics proposed. We examine a number of related problems, from containment and satisfiability to the interaction of SHACL with inference rules, and exhibit how different modellings of the language are useful for different problems. We also cover practical aspects of SHACL, discussing its implementations and state of adoption, to present a holistic review useful to practitioners and theoreticians alike.

1 Introduction

The Shapes Constraint Language (SHACL) [23] is a W3C recommendation language for the validation of RDF graphs. In SHACL, validation is based on *shapes*, which define particular constraints and specify which nodes in a graph should be validated against these constraints. The ability to validate data with respect to a set of constraints is of particular importance for RDF graphs, as they are schemaless by design. Validation can be used to detect problems in a dataset and it can provide data quality guarantees for the purpose of data exchange and interoperability. A set of constraints can also be interpreted as a "schema", functioning as one of the primary descriptors of a graph dataset, thus enhancing its understandability and usability. A set of SHACL shapes is called a *shapes graph*, but we refer to it as a SHACL *document* in order not to confuse it with the graphs that it is used to validate.

In this article we present a review of SHACL, which is composed of three main parts. In the first part, in Sects. 2 and 3, we review the SHACL specification. This part focuses on how shapes are defined, and how they are used for the purpose of validation. We highlight the main peculiarities of this language, and discuss how

© Springer Nature Switzerland AG 2022
M. Šimkus and I. Varzinczak (Eds.): Reasoning Web 2021, LNCS 13100, pp. 115–144, 2022.
https://doi.org/10.1007/978-3-030-95481-9_6

SHACL validation can be expressed either in terms of SPARQL queries, to facilitate its implementation, or in terms of *assignments* [13], to make it amenable to theoretical study. The syntax of SHACL is outside the scope of this review, and for the precise details on how to encode particular constraints we refer the reader to the SHACL specification [23]. We also do not discuss the process that lead to the development of SHACL, but it should be noted that this specification was built on top of a number of previous constraint languages, the most influential of which is Shape Expressions (ShEx) [48].

In the second part of our review, in Sects. 5 to 8, we present the formal properties of this language. This mainly revolves around a discussion of *recursion*. The semantics of recursion is not defined in the SHACL specification, and thus has been the subject of significant subsequent research [3]. The formal semantics of SHACL is given as a translation into SCL [37], a first order logic language that captures the entirety of the SHACL specification. Apart from validation, several standard decision problems are discussed, such as *satisfiability* and *containment*, along with an existing study on the interaction of SHACL with inference rules. We try to keep a consistent notation throughout this article and at times this notation might be different from the one in the original articles.

In Sect. 9, we review existing implementations of SHACL validators and their integration with mainstream graph databases. We also review prominent additional tools to manage SHACL documents, such as tools designed to automate or semi-automate the process of creating SHACL documents by exploiting graph data, ontologies, or other constraint languages. These approaches provide solutions to the cold start problem, and alleviate reliance on expert knowledge, which are typical problems of new technologies. We complement a discussion of these approaches with a review of prominent applications of SHACL in several domains; in summary, the abundance of SHACL related tools and applications highlights the remarkable level of maturity and adoption reached by this relatively new language.

2 Preliminaries

Before discussing SHACL, we briefly introduce our notation for RDF graphs [14]. With the term RDF *graph* (or just *graph*) we refer to a set of RDF *triples* (or just *triples*), where each triple $<s, p, o>$ identifies an edge with label p, called *predicate*, from a node s, called *subject*, to a node o, called *object*. Subjects, predicates and objects of RDF triples are collectively called RDF *terms*. The RDF terms that appear as subject and objects in the triples of a graph are called the *nodes* of the graph. Graphs in this article are represented in Turtle syntax using common XML namespaces, such as sh, rdf and rdfs to refer to, respectively, the SHACL, RDF, and RDFS [7] vocabularies. Queries over RDF graph will be expressed as SPARQL [42] queries.

In the RDF data model, subjects, predicates and objects are defined over different but overlapping domains. For example, while RDF terms of the IRI type can occupy any position in a triple, RDF terms of the *literal* type (representing

```
:EmployeeShape a sh:PropertyShape ;        :Anne a :Employee .
   sh:targetClass :Employee ;              :Bob a :Employee ;
   sh:path :hasOfficeNumber ;                 :hasOfficeNumber "18" ;
   sh:minCount 1 .                            :hasOfficeNumber "3" .
                                           :Carl a :Employee ;
                                              :hasOfficeNumber "171" .
                                           :David a :Customer .
```

Fig. 1. (Left) A sample SHACL document (shape graph) stating the constraint that every employee must have at least one office number. (Right) A sample RDF graph (data graph).

datatype values) can only appear in the object position. These differences are not central to the topics discussed in this review, and thus, for the sake of simplifying notation, we will assume that all elements of a triple are drawn from a single and infinite domain of constants. This corresponds to the notion of *generalized* RDF [14].

3   Overview of SHACL

The main application of SHACL is data validation. Data validation in SHACL requires two inputs: (1) an RDF graph G to be validated and (2) a SHACL document M that defines the conditions against which G must be evaluated. The SHACL specification defines the output of the data validation process as a *validation report*, detailing all the violations that were found in G of the conditions set by M. If the violation report contains no violations, a graph G is *valid* w.r.t. SHACL document M. The SHACL validation process can be abstracted into the following decision problem. Given a graph G and a SHACL document M, we denote with VALIDATE(G, M) the decision problem of deciding whether G is *valid* w.r.t. SHACL document M, that we call *validating* G against M.

For example, the graph on the left of Fig. 1 represents a SHACL document M_1, that defines the condition that every employee must have an office number. Therefore, the validation report for a graph and M_1 would list all of the instances of :Employee in the graph that do not have an office number. The validation report for M_1 and the data graph G_1 on the right of Fig. 1 contains a violation on node rdf:Anne, since she does not have an office number. Therefore G_1 is not valid w.r.t. M_1.

Formally, a SHACL *document* is a set of *shapes*. Validating a graph against a SHACL document involves validating it against each shape. Shapes restrict the structure of a valid graph by focusing on certain nodes and examining whether they satisfy their constraints. The main components of a shape are a *constraint* d and a *target definition* t. Constraints can be *evaluated* on any RDF node to determine whether that node *satisfies* or not the given constraints. A node that satisfies the constraint of a shape it is said to *conform* to that shape, or *not-conform* otherwise. If a shape has an empty constraint, all nodes trivially conform to the shape. Not all nodes of a graph must conform to all the shapes in the

SHACL document. The constraint definition of each shape defines which RDF nodes, called *target nodes*, must conform to that shape in order for the graph to be valid. A shape with an empty constraint definition does not have any target nodes. Through inter-shape referencing, as we will see below, additional nodes might be required to conform to certain shapes (or not, if negation is used) for the validation to succeed. Further irrelevant nodes within the graph do not play a role in validation of the shape, whether they conform to it or not. The SHACL document M_1 of our previous example, contains shape :EmployeeShape, whose constraint captures the property of "having an office number", and whose target definition targets only the RDF nodes of type Employee. Nodes of the Client type do not generate violations by not having an office number.

Formally, a shape is a tuple $\langle s, t, d \rangle$ defined by three components: (1) the shape name s, which uniquely identifies the shape; (2) the *target definition t*, and (3) the set of *constraints* which are used in conjunction, and hence hereafter referred to as the single constraint d. As demonstrated in Fig. 1, a SHACL document is itself an RDF graph. The graph representing a SHACL document is called a *shapes graph*, while the graph being validated is called a *data graph*. This approach to serialisation is similar to how OWL ontologies are serialised, and it serves a similar purpose. Thanks to this approach, a SHACL document does not require any dedicated infrastructure to be stored and shared. In fact, a SHACL document can be embedded directly into the very graph it validates, thus combining the shape graph and data graph into a single graph. Interestingly, with this serialisation, a shapes graph, being an RDF graph, can be itself subject to validation. The SHACL specification, in fact, defines a shapes graph that can be used to validate shapes graphs.

We will now look in more details at the two major components of shapes, namely target definitions and constraints.

3.1   SHACL Target Definitions

A SHACL target definition, within a constraint, is a set of *target declarations*. There are four types of target declarations defined in SHACL, each one taking an RDF term c as a parameter.

Node Targets A node target declaration on c targets that specific node.

Class-based Targets If a shape has a class-based target on c, then all the nodes in the graph that are of type (rdf:type) c are target nodes for that shape.

Subjects-of Targets If a shape has a subject-of target on c, then the target nodes for that shape are all the nodes in the graph that appear as subjects in triples with c as the predicate.

Objects-of Targets If a shape has an object-of target on c, then the target nodes for that shape are all the nodes in the graph that appear as objects in triples with c as the predicate.

The shape defined in Fig. 1 demonstrates an example of a class-based target targeting class :Employee. Similarly, to target a subject of a property, e.g., :worksAt, the second line of the shape definition would be substituted with:

```
sh:targetSubjectOf :worksAt.
```

Typically, a target declaration is used to select, among all the nodes in a graph, the ones to target for constraint validation. The *node target* declaration, however, behaves differently, as it targets a particular node regardless of whether this node occurs in the graph or not. An important implication of this is that empty graphs are not trivially valid, since node targets can detect violations on nodes external to the graph. If a target definition of a shape is empty, then that shape will have no target nodes. However, this does not mean that the constraint of that shape will not be evaluated on any nodes since, as mentioned, other shapes can refer to it and "pass it" a node to check for conformance.

3.2   Focus Nodes and Property Paths

When a target or another node is considered against a shape for conformity, we call it a *focus* node. Initially a shape focuses on its target nodes (these are the initial set of focus nodes). Additional focus nodes are obtained by following SHACL *property paths*, which we also refer to as just *paths*. SHACL property paths are a subset of SPARQL *property paths* and, as the name suggests, define paths in the RDF graph. The simplest type of path, called predicate path, corresponds to a single property IRI c. This path identifies all the nodes that are reachable in the RDF graph from the current focus node by following a single edge c. In other words, this path identifies all the RDF nodes in the object position of triples that have c as the predicate and the current focus node as the subject. More complex paths can be constructed by inverting the direction of a path, by concatenating two different paths one after the other, or by allowing the repetition of a path for a minimum, maximum or arbitrary number of times.

Based on the use of property paths, SHACL specification distinguishes shapes into two types: *node shapes* and *property shapes*. Intuitively, the constraint of a node shape is evaluated directly on the focus nodes of the shape. Instead, when using a property path, shapes must be declared as a property shapes. These are characterised by a path, and their constraints are evaluated over all of the nodes that can be reached from the focus nodes following such path. For example, the constraint that every employee's password must be at least 8 characters long can be represented by a property shape that targets employee nodes, and that has a relation such as :hasPassword as its path. In this way, the actual nodes that must satisfy the "at least 8 characters long" constraint are not the target nodes, but instead those that appear as objects in triples with an employee node as a subject, and :hasPassword as the predicate.

3.3   SHACL Constraints

The majority of the SHACL recommendation is dedicated to defining the different types of constraint components that can be used in SHACL constraints. The main type of constraint components are called *core constraint components*. These are the components that SHACL compliant systems typically support, and where most of the existing literature focuses on. The other main type of components are the SPARQL-*based constraint components*, that are used to embed SPARQL queries into SHACL constraints. This significantly increases the expressive power of such constraints. However, the inclusion of arbitrarily complex SPARQL queries can lead to performance issues, and can make such constraints harder to understand and use. It is also worth noting that, outside of the SHACL recommendation, a number of additional SHACL features[1] are currently being designed, and some of them might be included in further versions of SHACL. In the rest of this paper we will focus on core constraint components.

In order to better understand SHACL core constraint components, we propose a broad categorisation of these components into three main categories, depending on how they are evaluated on the focus nodes. Notice that most constraint components can be used in both node shapes and property shapes.

Graph Structure Components. These components define constraints that are evaluated at the level of triples of the graph, and focus on restrictions such as the minimum and maximum cardinality that the focus node must have for certain paths, or the RDF class that the focus node should be a type of. The shape defined in Fig. 1 demonstrates an example of a minimum cardinality constraint for predicate path :hasOfficeNumber. Two other salient constraints in this category are the property pair equality and disjointedness, that specify whether the two sets of nodes reachable from two different paths must be equal or disjoint, respectively.

Filter Components. These components define constraints that are evaluated at the level of nodes, and their evaluation is usually independent from the triples present in the graph. Filter constraints restrict the focus node (1) to be a particular RDF term, (2) to be of a particular type, such as IRI, blank node or literal, or (3) to be a literal that satisfies certain properties, such as being of the integer datatype, or a string produced by a certain regular expression.

Logical Components. Logical components define the standard logical operators of conjunction, disjunction and negation over other constraints.

While most core constraint components fall into one of these categories, the pair of constraints sh:lessThan and sh:lessThanOrEquals is a notable exception, as it is combines the properties of graph structure and filter components. These two constraints require all the nodes reachable by one path to be literals that are less than (resp. less than or equals) to the nodes reachable by a second path.

It is worth noting that all constraints but one, namely sh:closed, are not affected by triples with unknown predicates (i.e. predicates not occurring in the SHACL document). This means that if a graph is valid with respect to a set of those constraints, it would still remain valid if new triples with unknown predicates are added to the graph. Thus, given a non-empty graph G, valid w.r.t. a SHACL document M, graph $G \cup <s, p, o>$ is also valid w.r.t. M if (1) p does not occur in M and (2) M does not contain the sh:closed constraint component.

---

[1] https://w3c.github.io/shacl/shacl-af/ accessed on 18/6/21.

```
:EmployeeShapeB a sh:PropertyShape ;
    sh:targetClass :Employee ;
    sh:path :hasOfficeNumber ;
    sh:qualifiedMinCount 1 ;
    sh:qualifiedValueShape :OfficeNumberShape .

:OfficeNumberShape a sh:NodeShape ;
    sh:minLength 3 .
```

Fig. 2. A sample SHACL document stating the constraint that every employee must have at least one 3-characters or longer office number.

Intuitively, this means that those constraints restrict the usage of terms from a particular vocabulary, but they do not restrict in any way the graph from containing triples described using other vocabularies. The sh:closed component, on the other hand, restricts the predicates of the triples that have the focus node as a subject to belong to a predetermined finite set. Effectively, the sh:closed component can prohibit the use of unknown predicate relations for certain nodes in the graph, and thus prevent the inclusion of terms from other vocabularies. Interestingly, component sh:closed introduces an asymmety in SHACL, since it only affects triples where the focus node is the subject, and it is not possible to define a similar constraint for nodes in the object position.

A major feature of SHACL is that constraints can use the name of a shape to require a particular set of nodes to conform to that shape. This is called a *shape reference*. An example of a shape reference is demonstrated by the SHACL document in Fig. 2. This document contains shape :EmployeeShapeB which references shape :OfficeNumberShape. The former shape restricts all of its target nodes to having an edge :hasOfficeNumber to a node that conforms to the latter shape, having a string length of at least three characters. Validating the data graph in Fig. 1 with the SHACL document in Fig. 2 results in two violating nodes for shape :EmployeeShapeB. The first one is :Anne, who does not have an office number, and the second one is :Bob, whose office numbers all contain fewer than three digits.

Shape references can be recursive, that is, the constraint of a shape can reference the constraints of a second shape which, in turn, can reference the constraints of a third shape, and so on, creating a loop. Let S_0^d be the set of all the shape names occurring in a constraint d of a shape $\langle s, t, d \rangle$; these are the *directly* referenced shapes of s. Let S_{i+1}^d be the set of shapes in S_i^d union the directly referenced shapes of the constraints of the shapes in S_i^d.

Definition 1. *A shape $\langle s, t, d \rangle$ is recursive if $s \in S_\infty^d$; else it is non-recursive.*

Definition 2. *A SHACL document M is recursive if it contains a recursive shape, and non-recursive otherwise.*

The semantics of recursive SHACL documents are not defined in the SHACL specification. In Sect. 4 we review the official semantics of non-recursive SHACL documents, while in Sect. 5 we review the extended semantics for recursive SHACL document that have been proposed in the literature.

4 SHACL Validation

In this section we present the semantics of SHACL data validation, that is, the VALIDATE(G, M) decision problem, for any given graph G and SHACL document M. In Sect. 4.1 we review how validation is defined in the SHACL specification, with the help of SPARQL queries. While this query-based description of SHACL semantics can be easily translated into a concrete implementation, it does not lend itself well to theoretical investigation. In Sect. 4.2 we will discuss an alternative approach to defining SHACL semantics that is instead amenable to a formal study.

4.1 SHACL Validation by SPARQL Queries

The validation of an RDF graph G against a SHACL document M can be performed on a shape-by-shape basis. For each shape $\langle s, t, d \rangle$, this process involves verifying the fact that every node n, targeted by target definition t, satisfies constraint d. Intuitively, graph G is valid w.r.t. M if and only if this fact is true for every shape in M.

Given a graph G and a target definition t, the set of target nodes for t can be computed by evaluating a SPARQL query on G for each target declaration in t, and taking the union of the values returned by these queries. Table 1 details the corresponding SPARQL query for each of the four types of target declarations defined in SHACL. It should be noted that, by default, SHACL does not enforce any particular entailment regime. If an entailment regime is being adopted, then this should be taken into account when developing a SHACL validator. For example, if the RDFS entailment regime [7] is being considered, subclass inference should be accounted for when computing the set of entities of a given class. To accommodate for this entailment regime, the query for the node target in Table 1 could be updated to the following one.

```
SELECT ?x WHERE {
  ?x rdf:type/rdfs:subClassOf* c
}
```

Once an RDF term has been identified as being in the target of a shape, evaluating whether it conforms to the shape can be done using SPARQL queries. In the SHACL specification, in fact, several core constraint components are defined with respect to SPARQL queries. Most notably, the semantics of SHACL filter components is in direct dependence to the semantics of SPARQL filter functions. For example, the sh:minLength constraint component restricts a focus node to having a string length equal or larger than a given number. Formally, a focus node n has a sh:minLength of j if and only if the following SPARQL query evaluates to true.

Table 1. Target declarations and their corresponding SPARQL queries to compute the set of target nodes on a given graph

Target declaration	SPARQL query
Node target (node c)	SELECT ?x WHERE { VALUES ?x { c } }
Class target (class c)	SELECT ?x WHERE { ?x rdf:type c. }
Subjects-of target (predicate c)	SELECT ?x WHERE { ?x c ?y. }
Objects-of target (relation c)	SELECT ?x WHERE { ?y c ?x. }

```
ASK {
  FILTER (STRLEN(str(n)) >= j) .
}
```

Not all SHACL constraints, however, can be easily verified by a single SPARQL query. Evaluating whether a constraint that contains shape references is satisfied by a focus node, in fact, might involve evaluating whether other constraints are satisfied by other nodes which, in turn, might require even further constraint evaluations. For example, in order to evaluate whether node rdf:Carl from the data graph in Fig. 1 conforms to shape rdf:EmployeeShapeB from Fig. 2, we would need to evaluate whether his office number, namely RDF term "171", conforms to shape rdf:OfficeNumberShape. This is especially problematic in case of recursion, as it could generate an infinite series of constraint evaluations. For non-recursive SHACL documents, Corman et al. [12] showed that it is always possible to check the validity of a graph using a single SPARQL query. For example, a graph can be checked against the SHACL document of Fig. 2 by evaluating the following SPARQL query.

```
SELECT ?x WHERE {
  ?x a :Employee .
  FILTER NOT EXISTS {
    ?x :hasOfficeNumber ?y .
    FILTER (STRLEN(str(?y)) >= 3) .
  }
}
```

This query selects all RDF nodes of type Employee that do not have an office number with at least three characters. Thus, any RDF term returned by this query is a node violating a shape of the SHACL document. If this query evaluates to an empty set, then the graph that it is evaluated on is valid with respect to the SHACL document.

4.2 Shape Assignments: A Tool for Defining SHACL Validation

The SPARQL-based approach to SHACL validation does not provide a concise and formal description of SHACL semantics. Moreover, it does not provide us with a terminating procedure to check graphs in the face of SHACL recursion. In this section we review the concept of *shape assignments* (or just *assignments*) [13], which can be used to address the above mentioned problems.

As defined in Table 1, a target declaration t is a unary query over a graph G. We denote with $G \models t(n)$ that a node n is *in the target* of t with respect to a graph G. If t is empty, no node in any graph is in the target of t. The definition of whether a node conforms to a shape, as we previously discussed, does not only depend on the graph G, but it might also depend, due to shape references, on whether other nodes conform to other shapes. Intuitively, the concept of *assignments* [13] is used to keep track, for every RDF node, of all the shapes that it conforms to, and all of those that it does not. Given a document M and a graph G, we denote nodes(G, M) the set of nodes in G together with any extra ones referenced by the node target declarations in M. With shapes(M) we refer to all the shape names in a document M.

Definition 3. *Given a graph G, and a SHACL document M, an assignment σ for G and M is a function mapping nodes in* nodes(G, M), *to subsets of* shapes$(M) \cup \{\neg s \mid s \in$ shapes$(M)\}$, *such that for all nodes n and shape names s, $\sigma(n)$ does not contain both s and $\neg s$.*

Expression $[\![d]\!]^{n,G,\sigma}$ denotes the evaluation of constraint d on a node n w.r.t. a graph G under an assignment σ, as defined in [13]. If $[\![d]\!]^{n,G,\sigma}$ is True (resp. False) we say that node n satisfies (resp. does not satisfy) constraint d w.r.t. G under σ. For any graph G and assignment σ, fact $s \in \sigma(n)$ (resp. $\neg s \in \sigma(n)$) denotes the fact that node n conforms (resp. does not conform) to s w.r.t. G under σ. Expression $[\![d]\!]^{n,G,\sigma}$ evaluates to True, False or Undefined values of Kleene's 3-valued logic, and the truth value of any shape reference in d is computed using the assignment (it should be noted that the Undefined value never occurs in non-recursive shapes, but it is used to define possible extended semantics in the face of recursion). In other words, whenever a truth value in the evaluation of $[\![d]\!]^{n,G,\sigma}$ depends on whether another node j conforms to a shape s', with constraints d', this is not resolved by evaluating $[\![d']\!]^{j,G,\sigma}$, but instead it is True if $s' \in \sigma(j)$, False if $\neg s' \in \sigma(j)$, or else Undefined. This, in turn, eliminates the problem of a potentially infinite series of constraint evaluations.

The semantics of SHACL validation can be defined with respect to a particular type of assignments, called *faithful* [13].

Definition 4. *For all graphs G, SHACL documents M and assignments σ, assignment σ is faithful w.r.t. G and M, denoted with $(G, \sigma) \models M$, if the following two conditions hold for any shape $\langle s, t, d \rangle$ in* shapes(M) *and node n in* nodes(G, M):

(1) $s \in \sigma(n)$ iff $[\![d]\!]^{n,G,\sigma}$ is True; and $\neg s \in \sigma(n)$ iff $[\![d]\!]^{n,G,\sigma}$ is False;
(2) if $G \models t(n)$ then $s \in \sigma(n)$.

Condition (1) ensures that the facts denoted by the assignment are correct; while condition (2) ensures that the assignment is compatible with the target definitions. Condition (2) is trivially satisfied for SHACL documents where all target definitions are empty. Later we will want to discuss assignments where the first property of Definition 4 holds, but not necessarily the second, in order to reason about the existence of alternative assignments that are correct (as in, they satisfy the first part of Definition 4) but that are not faithful. In fact, these will be faithful assignments to a document that is "stripped empty" of target definitions. Let $M^{\backslash t}$ denote the SHACL document obtained from substituting all target definitions in SHACL document M with the empty set. The following lemma holds:

Lemma 1. *For all graphs G, SHACL documents M and assignments σ, condition (1) from Definition 4 holds for any shape s in* shapes(M) *and node n in* nodes(G, M) *iff* $(G, \sigma) \models M^{\backslash t}$.

The existence of a faithful assignment is a necessary and sufficient condition for validation for non-recursive SHACL documents [13]. As we will see later, this is also necessary condition for all the other extended semantics.

Definition 5. *A graph G is valid w.r.t. a non-recursive SHACL document M if there exists an assignment σ such that $(G, \sigma) \models M$.*

5   SHACL Recursion

The semantics of recursion in SHACL documents is left undefined in the SHACL specification [23], and this gives rise to several possible interpretations. In this section we consider *extended* semantics of SHACL that define how to validate graphs against recursive SHACL documents. We focus on existing extended semantics that follow monotone reasoning. These can be characterised by two dimensions, namely the choice between *partial* and *total* assignments [13] and between *brave* and *cautious* validation [3], which we will subsequently define. Put together, these two dimensions define the four extended semantics of *brave-partial*, *brave-total*, *cautious-partial* and *cautious-total*. We will not go into the details of the less obvious dimension of *stable-model* semantics [3], which relates SHACL to non-monotone reasoning in logic programs.

As mentioned in the previous section, assignments can specify a truth value of True, False or Undefined to whether a node conforms to given shape. The truth value of Undefined, which does not occur in non-recursive SHACL documents, can instead play an important role in validating SHACL under recursion. Intuitively, this happens during validation, when recursion makes it impossible for a node n to either conform or not to conform to a shape s but, at the same time, validity does not depend on whether n conforms to shape s or not. Consider for example the following SHACL document, containing a single shape $\langle s^*, \emptyset, d^* \rangle$ (with name :InconsistentS in this example). This shape is defined as the negation of itself, that is, given a node n, a graph G and an assignment σ, fact $[\![d^*]\!]^{n,G,\sigma}$ is true iff $\neg s^* \in \sigma(n)$, and false iff $s^* \in \sigma(n)$.

```
:InconsistentS a sh:NodeShape ;
   sh:not :InconsistentS .
```

It is easy to see that any assignment that maps a node to either s^* or $\neg s^*$ is not faithful, as it would violate condition (1) of Definition 4. However, an assignment that maps every node of a graph to the empty set would be faithful for that graph and document $\{s^*\}$. Intuitively, this means that nodes in the graph cannot conform nor not conform to shape s^*, but since this shape does not have any target node to validate, then the graph can still be valid. The fact of whether nodes conform or not conform to shape s^* can thus be left as "undefined".

This type of validation, for recursive SHACL documents, is called validation with partial assignments. More specifically, validation under brave-partial semantics simply extends the criterion of Definition 5 to recursive SHACL documents. All other extended semantics are constructed by adding additional conditions to brave-partial semantics. The term "partial" should not be interpreted as the fact that it describes only "part" of nodes of a graph, or that it describes the relationship of a node to only "part" of the shapes. Within a partial assignment, the conformance of every node to every shape is precisely specified by one of three truth values, and the term "partial" only indicates that one of these three truth values is Undefined.

Definition 6. *A graph G is valid w.r.t. a* SHACL *document M under brave-partial semantics if there exists an assignment σ such that $(G, \sigma) \models M$.*

In the SHACL specification, nodes either conform to, or not conform to a given shape, and the concept of an "undefined" level of conformance is arguably alien to the specification. It is natural, therefore, to consider restricting the evaluation of a constraint to the True and False values of boolean logic. This is achieved by restricting assignments to be *total*.

Definition 7. *An assignment σ is total w.r.t. a graph G and a* SHACL *document M if, for all nodes n in* nodes(G, M) *and shapes $\langle s, t, d \rangle$ in M, either $s \in \sigma(n)$ or $\neg s \in \sigma(n)$.*

For any graph G and SHACL document M we denote with $A^{G,M}$ and $A_T^{G,M}$, respectively, the set of assignments, and the set of total assignments for G and M. Trivially, $A_T^{G,M} \subseteq A^{G,M}$ holds.

Definition 8. *A graph G is valid w.r.t. a* SHACL *document M under brave-total semantics if there exists an assignment σ in $A_T^{G,M}$ such that $(G, \sigma) \models M$.*

Since total assignments are a more specific type of assignments, if a graph G is valid w.r.t. a SHACL document M under brave-total semantics, than it is also valid w.r.t. M under brave-partial semantics. The reverse, instead, is only true for non-recursive SHACL documents. In fact, as shown in [13], if there exists a faithful assignment for a graph G and a non-recursive document M, then there exists also a total faithful assignment for G and M. Therefore, the definition of validity under brave-total semantics (Definition 8), for non-recursive SHACL documents, coincides with the standard definition of validation (Definition 5).

While total assignments can be seen as a more natural way of interpreting the SHACL specification, they are not without issues when recursive SHACL documents are considered. Going back to our previous example, we can notice that there cannot exist a total faithful assignment for the SHACL document containing shape :InconsistentS, for any non-empty graph. This is a trivial consequence of the fact that no node can conform to, nor not conform to, shape :InconsistentS. This, however, is in contradiction with the SHACL specification, which implies that a SHACL document without target declarations in any of its shapes (such as the one in our example) should trivially validate any graph. If there are no target declarations, in fact, there are no target nodes on which to verify the conformance of certain shapes, and thus no violations should be detected.

The second and last dimension that we consider is the difference between brave and cautious validation. When a SHACL document M is recursive, there might exist multiple assignments σ satisfying property (1) of Definition 4, that is, such that $(G, \sigma) \models M^{\backslash t}$. Intuitively, these can be seen as equally "correct" assignments with respect to the constraints of the shapes, and brave validation only checks whether at least one of them is compatible with the target definitions of the shapes. Cautious validation, instead, represents a stronger form of validation, where all such assignments must be compatible with the target definitions.

Definition 9. *A graph G is valid w.r.t. a* SHACL *document M under* cautious-partial *(resp.* cautious-total*) semantics if it is (1) valid under* brave-partial *(resp.* brave-total*) semantics and (2) for all assignments σ in $A^{G,M}$ (resp. $A_T^{G,M}$), it is true that if $(G, \sigma) \models M^{\backslash t}$ holds then $(G, \sigma) \models M$ also holds.*

To exemplify this distinction, consider the following SHACL document M_1. This document requires the daily special of a restaurant, node :DailySpecial, to be vegetarian, that is, to conform to shape :VegDishShape. This shape is recursively defined as follows. Something is a vegetarian dish if it contains an ingredient, and all of its ingredients are vegetarian, that is, entities conforming to the :VegIngredientShape. A vegetarian ingredient, in turn, is an ingredient of at least one vegetarian dish.

```
:VegDishShape a sh:PropertyShape ;
   sh:targetNode :DailySpecial ;
   sh:path :hasIngredient ;
   sh:minCount 1 ;
   sh:qualifiedMaxCount 0 ;
   sh:qualifiedValueShape [ sh: not :VegIngredientShape ] .

:VegIngredientShape a sh:PropertyShape ;
   sh:path [ sh:inversePath :hasIngredient ] ;
   sh:node :VegDishShape .
```

Consider now a graph G_1 containing the following triple.

```
:DailySpecial :hasIngredient :Chicken .
```

Due to the recursive definition of :VegDishShape, there exist two different assignments σ_1 and σ_2, which are both faithful for G_1 and $M_1^{\backslash t}$. In σ_1, no node in G_1 conforms to any shape, while σ_2 differs from σ_1 in that node :DailySpecial conforms to :VegDishShape and node :Chicken conforms to :VegIngredientShape. Essentially, either both the dish and the ingredient from graph G_1 are vegetarian, or neither is. Therefore, σ_2 is faithful for G_1 and M_1, while σ_1 is not. The question of whether the daily special is a vegetarian dish or not can be approached with different levels of "caution". Under brave validation, graph G_1 is valid w.r.t. M_1, since it is possible that the daily special is vegetarian. Cautious validation, instead, takes the more conservative approach, and under its definition G_1 is not valid w.r.t. M_1, since it is also possible that the daily special is not vegetarian. When analysing such recursive definitions, one might want to exclude "unfounded" assignments, that is, assignments that assign certain shapes to a node for no other reason than to allow the validation of a graph. This is achieved by the recursive semantics for SHACL proposed in [3], which is based on the concept of *stable models* from Answer Set Programming.

For each extended semantics, the definition of validity of a graph G with respect to a SHACL document M, denoted by $G \models M$, is summarised in the following list.

brave-partial there exists an assignment that is faithful w.r.t. G and M;
brave-total there exists an assignment that is total and faithful w.r.t. G and M;
cautious-partial there exists an assignment that is faithful w.r.t. G and M, and every assignment that is faithful w.r.t. G and $M^{\backslash t}$ is also faithful w.r.t. G and M.
cautious-total there exists an assignment that is total and faithful w.r.t. G and M, and every assignment that is total and faithful w.r.t. G and $M^{\backslash t}$ is also faithful w.r.t. G and M.

6   Formal Languages for SHACL

In this section we review the two main formal languages that have been proposed to model the semantics of SHACL. We first discuss a complete first-order formalisation of SHACL, which can be used to study a number of decision problems. We then present a simplified language that effectively models SHACL constraints for the purpose of validation.

6.1   SCL, a First-Order Language for SHACL

In order to formally study SHACL, it is convenient to abstract away from the syntax of its RDF and SPARQL representations. The SCL first order language [36,37] is currently the only complete formalisation of SHACL into a formal logical system. The expressiveness of this language covers all of the SHACL target declarations and all of the SHACL core constraint components, including the filter components, which are less commonly studied. This language captures the semantics of whole SHACL documents, and it can be used to study a number of related decision problems, including validation. The relation between SHACL and SCL is given by translation τ [36], such that, given a SHACL document M, the first order sentence $\tau(M)$ is the translation of M into SCL. We identify the inverse translation with τ^-.

Before defining SCL and its properties, we must define how RDF graphs and assignments are modelled in this logical framework. The domain of discourse is assumed to be the set of RDF terms. Triples are modelled as binary relations, with atom $R(s,o)$ corresponding to triple $<s,R,o>$. A minus sign identifies the *inverse* role, i.e. $R^-(s,o) = R(o,s)$. Binary relation name isA represents class membership triples $<s,\texttt{rdf:type},o>$ as $\texttt{isA}(s,o)$. Assignments are modelled with a set of monadic relations Σ, called *shape relations*. Each SHACL shape s is associated with a unique shape relation $\Sigma_\mathbf{s}$ in SCL. Facts $\Sigma(x)$ (resp. $\neg\Sigma(x)$) describe an assignment σ such that $\mathbf{s} \in \sigma(x)$ (resp. $\neg\mathbf{s} \in \sigma(x)$). Since this logical framework adopts boolean logic, $\forall x.\ \Sigma(x) \vee \neg\Sigma(x)$ holds, by the law of excluded middle. Thus shape relations define total assignments.

Given a graph G and an assignment σ, we now define their respective translations G^τ and σ^τ into first order structures.

Definition 10. *Given a graph G, fact $p(s,o)$ is true in the first order structure G^τ iff $<s,p,o> \in G$.*

Definition 11. *Given a total assignment σ, fact $\Sigma_s(n)$ is true in the first order structure σ^τ iff $\mathbf{s} \in \sigma(n)$.*

Definition 12. *Given a graph G and a total assignment σ, the first order structure I induced by G and σ is the disjoint union of structures G^τ and σ^τ. Given a first order structure I: (1) the graph G induced by I is the graph that contains triple $<s,p,o>$ iff $I \models p(s,o)$ and (2) the assignment σ induced by I is the assignment such that, for all nodes n and shape relations Σ_s, fact $\mathbf{s} \in \sigma(n)$ is true iff $I \models \Sigma_s(n)$ and $\neg\mathbf{s} \in \sigma(n)$ iff $I \not\models \Sigma_s(n)$.*

The existence of faithful assignments using SCL and its standard model-theoretic semantics is presented in the following theorem [37]. Trivially, this also defines what condition, in SCL, corresponds to validation under the brave-total extended semantics (Definition 8), which also defines validation for non-recursive SHACL documents (Definition 5).

Theorem 1. *For any graph G, total assignment σ and SHACL document M, it is true that $(G,\sigma) \models M$ iff $I \models \tau(M)$, where I is the first order structure induced by G and σ.*

For any first order structure I and SCL formula ϕ, it is true $I \models \phi$ iff $(G,\sigma) \models \tau^-(\phi)$, where G and σ are, respectively, the graph and assignment induced by I.

Sentences in the SCL language follow the φ grammar in Definition 13.

Definition 13. *The SHACL first order language (SCL, for short) is the set of first order sentences built according to the following context-free grammar, where*

Table 2. Translation of a shape with name s with a target definition t, into an SCL target axiom.

Target declaration in t	SCL target axiom
Node target (node c)	$\Sigma_s(c)$
Class target (class c)	$\forall x.\mathtt{isA}(x,c) \to \Sigma_s(x)$
Subjects-of target (relation R)	$\forall x,y.R(x,y) \to \Sigma_s(x)$
Objects-of target (relation R)	$\forall x,y.R^-(x,y) \to \Sigma_s(x)$

c *is a constant from the domain of* RDF *terms*, Σ *is a shape relation*, F *is a filter relation, with shape relations disjoint from filter relations*, R *is a binary-relation name*, $\star$ *indicates the transitive closure of the relation induced by* $\pi(x,y)$, *the superscript* $\pm$ *refers to a relation or its inverse, and* $n \in \mathbb{N}$.

$$\varphi := \top \mid \varphi \wedge \varphi$$
$$\mid \Sigma(c) \mid \forall x.\, \mathit{isA}(x,c) \to \Sigma(x) \mid \forall x,y.\, R^{\pm}(x,y) \to \Sigma(x)$$
$$\mid \forall x.\, \Sigma(x) \leftrightarrow \psi(x);$$
$$\psi(x) := \top \mid \neg\psi(x) \mid \psi(x) \wedge \psi(x) \mid x = c \mid F(x) \mid \Sigma(x) \mid \exists y.\, \pi(x,y) \wedge \psi(y)$$

$$\mid \neg\exists y.\, \pi(x,y) \wedge R(x,y) \qquad\qquad\qquad\qquad\qquad\qquad\qquad [D]$$
$$\mid \forall y.\, \pi(x,y) \leftrightarrow R(x,y) \qquad\qquad\qquad\qquad\qquad\qquad\qquad [E]$$
$$\mid \forall y,z.\, \pi(x,y) \wedge R(x,z) \to \varsigma(y,z) \qquad\qquad\qquad\qquad\quad [O]$$
$$\mid \exists^{\geq n} y.\, \pi(x,y) \wedge \psi(y); \qquad\qquad\qquad\qquad\qquad\qquad\qquad [C]$$

$$\pi(x,y) := R^{\pm}(x,y)$$
$$\mid \exists z.\, \pi(x,z) \wedge \pi(z,y) \qquad\qquad\qquad\qquad\qquad\qquad\qquad [S]$$
$$\mid x = y \vee \pi(x,y) \qquad\qquad\qquad\qquad\qquad\qquad\qquad\qquad [Z]$$
$$\mid \pi(x,y) \vee \pi(x,y) \qquad\qquad\qquad\qquad\qquad\qquad\qquad\qquad [A]$$
$$\mid (\pi(x,y))^{\star}; \qquad\qquad\qquad\qquad\qquad\qquad\qquad\qquad\qquad [T]$$

$$\varsigma(x,y) := x <^{\pm} y \mid x \leq^{\pm} y.$$

Symbol φ corresponds to a SHACL document. An SCL sentence could be empty ($\top$), a conjunction of documents, a *target axiom* representing a target definition (a production of the 3rd, 4th and 5th production rule) or a *constraint axiom* representing a constraint (a production of the last production rule). Target axioms take one of three forms, based on the type of target declarations. The translation of SHACL target declarations into SCL target axioms is summarised in Table 2. Letters in square brackets are annotations for naming SCL components and thus are not part of the grammar. These letters are essentially first-letter abbreviations of *prominent* SHACL components (that together define fragments of SCL), and are also listed in Table 3.

The non terminal symbol $\psi(x)$ corresponds to the subgrammar of the SHACL constraints components. Within this subgrammar, $\top$ identifies an empty con-

Table 3. Relation between prominent SHACL components and SCL expressions.

Abbr.	Name	SHACL component	Corresponding expression
S	Sequence Paths	Sequence Paths	$\exists z . \pi(x,z) \wedge \pi(z,y)$
Z	Zero-or-one Paths	sh:zeroOrOnePath	$x = y \vee \pi(x,y)$
A	Alternative Paths	sh:alternativePath	$\pi(x,y) \vee \pi(x,y)$
T	Transitive Paths	sh:zeroOrMorePath sh:oneOrMorePath	$(\pi(x,y))^*$
D	Property Pair Disjointness	sh:disjoint	$\neg \exists y.\pi(x,y) \wedge R(x,y)$
E	Property Pair Equality	sh:equals	$\forall y . \pi(x,y) \leftrightarrow R(x,y)$
O	Property Pair Order	sh:lessThan sh:lessThanOrEquals	$x \leq^{\pm} y$ and $x <^{\pm} y$
C	Cardinality Constraints	sh:qualifiedValueShape sh:qualifiedMinCount sh:qualifiedMaxCount	$\exists^{\geq n} y . \pi(x,y) \wedge \psi(y)$ with $n \neq 1$

straint, $x = c$ a constant equivalence constraint and F a monadic filter relation (e.g. $F^{\mathrm{IRI}}(x)$, true iff x is an IRI). Filters components are captured by $F(x)$ and the O component. The C component captures qualified value shape cardinality constraints. The E, D and O components capture the equality, disjointedness and order property pair components. The $\pi(x,y)$ subgrammar models SHACL property paths. Within this subgrammar S denotes sequence paths, A denotes alternate paths, Z denotes a zero-or-one path and T denotes a zero-or-more path.

Translation τ results in a subset of SCL formulas, called *well-formed* defined subsequently, and the inverse translation τ^- only takes well formed sentences as an input. An SCL formula ϕ is well-formed iff for every shape relation Σ, formula ϕ contains exactly one constraint axiom with relation Σ on the left-hand side of the implication. Intuitively, this condition ensures that every shape relation is "defined" by a corresponding constraint axiom. The translation of the document from Fig. 2, into a well-formed SCL sentence, via τ, is the following. Arguably, this logic notation might seem easier to read and understand than the SHACL syntax of Fig. 2.

$$\left(\forall x. \ \mathtt{isA}(x, \mathtt{:Employee}) \rightarrow \Sigma_{\mathtt{:EmployeeShapeB}}(x)\right)$$

$$\wedge \left(\forall x. \ \Sigma_{\mathtt{:EmployeeShapeB}}(x) \leftrightarrow \exists y. \ R_{\mathtt{:hasOfficeNumber}}(x,y) \wedge \Sigma_{\mathtt{:OfficeNumberShape}}(y)\right)$$

$$\wedge \left(\forall x. \ \Sigma_{\mathtt{:OfficeNumberShape}}(x) \leftrightarrow F^{\mathrm{length} \geq 3}(y)\right)$$

The language defined without any of these constructs is called the *base* language, denoted $\varnothing$. On top of the base language different syntactic fragments of SCL are defined by considering different combinations of features allowed. We name these fragments by concatenating the letters that represent the features allowed, into a single name. For example, SA identifies the fragment that only allows the base language, sequence paths and alternate paths. This means that in order to write an SCL document in SA, one can only use the production rules of Definition 13 that are not annotated with any feature (base language) or those identified by abbreviations S and A.

The SHACL specification presents an unusual asymmetry in the fact that equality, disjointedness and order components force one of their two path expressions to be an atomic relation. This can result in situations where the order constraints can be defined in just one direction, since only the less-than and less-than-or-equal property pair constraints are defined in SHACL. The O fragment models a more natural order comparison that includes the $>$ and $\geq$ components. The fragment where the order relations in the $\varsigma(x, y)$ subgrammar cannot be inverted is denoted O'.

When interpreting an SCL sentence, particular care should be paid to the semantics of filter relation. The interpretation of each filter relation, such as $F^{\mathrm{IRI}}(x)$, is the subset of the domain of discourse on which the filter is true. This interpretation is constant across all models, and defines the semantics of the filter. When considering the decision problem of validation, filter relations in SCL must be suitably defined by interpreted relations (similarly to how the equality operator is). When considering additional decision problems, such as satisfiability and containment (which will be discussed in Sect. 7), the semantics of filters can be axiomatisatised, thus removing the need for special interpreted relations. The filter axiomatisation presented in [37] captures the semantics of all SHACL filters with the single exception of sh:pattern, as this filter defines complex non-standard regular expressions based on the SPARQL REGEX function [42].

6.2 $\mathcal{L}$, a Language for SHACL Constraint Validation

Another major language used to study SHACL is $\mathcal{L}$ which was presented in [13] and paved the way to subsequent formal studies of SHACL. The $\mathcal{L}$ language differs from SCL in scope and purpose. While SCL sentences describe whole SHACL documents, sentences in $\mathcal{L}$ describe individual SHACL constraints. The $\mathcal{L}$ language is primarily designed to investigate the complexity of SHACL validation. As such, it relies on assumptions that do not hold when studying other decision problems such as satisfiability and containment, which, instead, can be studied using SCL. In particular, $\mathcal{L}$ assumes that all filter components can be evaluated on a node in constant time, and thus are all equivalent, for the purposes of validation. Thanks to this reduced scope, $\mathcal{L}$ seems less complex than SCL, and it is a useful formalism to study the evaluation of SHACL constraints. The semantics of an $\mathcal{L}$ sentence ϕ is defined in [13] through the use of faithful assignments. In particular, [13] fixes a lookup table that provides the truth value of the evaluation of ϕ on a node n for a graph G and an assignment σ. Instead, SCL relies on the standard model-theoretic semantics.

The grammar of $\mathcal{L}$ sentences is given next. In this grammar s is a shape name; I is an IRI; r is a SHACL property path; n is a positive integer.

$$\phi := \top \mid s \mid I \mid \phi_1 \wedge \phi_2 \mid \neg\phi \mid \geq_n r.\phi \mid \mathrm{EQ}(r_1, r_2)$$

Table 4. Correspondence between an $\mathcal{L}$ sentence ϕ, and SCL $\psi^\phi(x)$ expressions, such that a constraint ϕ is satisfied on a node n w.r.t. a graph G and an assignment σ iff $I \models \psi^\phi(n)$, where I is the first order structure induced by G and σ. It is assumed that paths are expressed using the $\pi(x,y)$ subgrammar of SCL, and that r_2 is an IRI.

$\mathcal{L}$ expression ϕ	Corresponding SCL $\psi^\phi(x)$
$\top$	$\top$
s	$\Sigma_s(x)$
I	$x = I$
$\phi_1 \wedge \phi_2$	$\psi^{\phi_1}(x) \wedge \psi^{\phi_2}(x)$
$\neg\phi$	$\neg\psi^\phi(x)$
$\geq_n r.\phi$	$\exists^{\geq n} y . r(x,y) \wedge \psi^\phi(y)$
$EQ(r_1, r_2)$	$\forall y . r_1(x,y) \leftrightarrow r_2(x,y)$

Table 4 defines the correspondence between $\mathcal{L}$ and the $\psi(x)$ sub-grammar of SCL. It is easy to see that $\mathcal{L}$ sentences correspond to a subset of the $\psi(x)$ sub-grammar of SCL, assuming that r_2 denotes a predicate path. This assumption is required as in $\mathcal{L}$ both arguments of $EQ(r_1, r_2)$, which captures the SHACL equality operator (sh:equals), are path expressions. This is a generalisation of SHACL, since the SHACL specification requires one of the two paths to be a simple predicate path, or in other words, an IRI. It should also be noted that $\mathcal{L}$ does not model property pair order components (denoted O in SCL), and that the sh:closed component is modelled using path expression operators not supported by SHACL paths. The SHACL disjoint constraint component (denoted D in SCL) is only implicitly included in $\mathcal{L}$ when considering recursion. It is possible, in fact, to represent a disjoint constraint component in $\mathcal{L}$ using two auxiliary recursive shapes [13].

7   SHACL Decision Problems

Several existing pieces of work in the literature focus on SHACL, and several related decision problems have been investigated. In Sect. 7.1 we review existing work on the core decision problem for SHACL, namely validation. Unlike validation, which studies the relationship between a SHACL document and an RDF graph, the decision problems of *satisfiability* and *containment*, reviewed in Sect. 7.2, focus on intrinsic properties of SHACL documents and their components.

7.1   Validation

Validation is a core decision problem for SHACL, since the main application of this language is the validation of RDF graphs. This decision problem is decidable for all of the semantics discussed in this article, including the four extended

Table 5. Data complexity of SHACL validation, results from [12].

Fragment	Data complexity of validation
SHACL$^{non-rec}$	NL-c
SHACL^{+}	PTIME-c
SHACL^{rec}	NP-c

semantics. The complexity lower bounds for validation, however, depend on the fragment of SHACL being considered. Table 5 lists the data complexity of three fragments of SHACL given in [3,13]. The three fragments are (1) SHACL$^{non-rec}$, the fragment of non-recursive SHACL documents built using $\mathcal{L}$ constraints; (2) SHACL^{+}, the fragment of SHACL documents built using $\mathcal{L}$ constraints with a restricted use of negation, that is, substituting the $\neg\phi$ production rule of $\mathcal{L}$ into $\phi_1 \vee \phi_2$; and (3) SHACL^{rec}, the fragment of SHACL documents built using $\mathcal{L}$ constraints. The most expressive of these fragments, SHACL^{rec}, is NP-complete in data complexity.

7.2 Satisfiability and Containment

Satisfiability and containment are standard decision problems that have been investigated in the context of SHACL. These two decision problems, unlike validation, do not take a graph as an input. Instead, they focus on SHACL documents, shapes or constraints. Given any notion of validity from one of the semantics defined earlier, the following decision problems are defined. For simplicity, when discussing satistiability and containment, we will assume the use of the semantics of validation from Definitions 8 and 5.

Definition 14. *A SHACL document M is satisfiable iff there exists a graph G such that $G \models M$. Deciding whether a SHACL document is satisfiable is the decision problem of SHACL satisfiability.*

Definition 15. **SHACL** *Containment: For all SHACL documents M_1, M_2, we say that M_1 is contained in M_2, denoted $M_1 \subseteq M_2$, iff for all graphs G, if $G \models M_1$ then $G \models M_2$. Deciding whether a SHACL document is contained in another is the decision problem of SHACL containment.*

Two SHACL documents M_1 and M_2 that are contained in each other ($M_1 \subseteq M_2$ and $M_2 \subseteq M_1$) are *semantically equivalent*. Two semantically equivalent documents are not necessarily equivalent syntactically, since in SHACL the same constraint can be expressed using different sets of shapes.

The satisfiability and containment decision problems for SHACL can be polynomially reduced to the satisfiability decision problem for SCL, defined as follows in the natural way [37].

Definition 16. *An SCL sentence ϕ is satisfiable iff there exists structure Ω such that $\Omega \models \phi$. Deciding whether a SCL sentence is satisfiable is the decision problem of SCL satisfiability.*

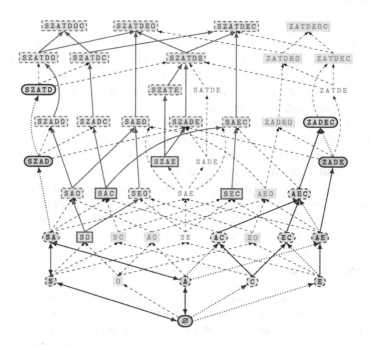

Fig. 3. [37] Decidability and complexity map of SCL satisfiability. Round (blue) and square (red) nodes denote decidable and undecidable fragments, respectively. Solid borders on nodes correspond to theorems in this paper, while dashed borders are implied results. Directed edges indicate inclusion of fragments, while bidirectional edges denote polynomial-time reducibility. Solid edges are preferred derivations to obtain tight results, while dotted ones leads to worst upper-bounds or model-theoretic properties. Finally, a light blue background indicates that the fragment enjoys the finite-model property, while those with a light red background do not satisfy this property. (Color figure online)

This reduction allows us to study the decidability and complexity of the SHACL satisfiability and containment problems for a given SHACL fragment by studying the decidability and complexity of SCL satisfiability, for the corresponding fragments. The results of this study, published in [37], are summarised in Fig. 3. Negative results indicate the undecidability of both the SCL fragment, and the corresponding SHACL fragment. Positive results, shown in round blue in the figure, indicate that both satisfiability and containment are decidable, for that fragment of non-recursive SHACL, and are accompanied with complexity upper-bounds. Starting from the negative results, SHACL satisfiability and containment is, in general, undecidable. This was shown even for several non-recursive fragments, through a semi-conservative reduction from the standard domino problem [5,39,50], which is an undecidable decision problem. More specifically, the SHACL satisfiability problems for the S O, S A C, S E C, S E O', and S Z A E fragments are undecidable [37].

Positive results are obtained by noticing that several SCL fragments are included in decidable fragments of first order logic. For example, the S Z A T D fragment of SCL is included in the extension of the unary-negation fragment of first-order logic with arbitrary transitive relations, which can be solved in 2Exp-Time [1,15,20]. The complexity upper bounds identified in [37] for SHACL fragments range from ExpTime to 2ExpTime. Both decision problems are defined over SHACL documents which, similarly to the schema of a dataset, could be assumed to be of small or constant size.

Up until this point we considered the satisfiability and containment problems defined at the level of SHACL documents. However, it is possible to study variations of these problems at different levels of granularity. For example, the satisfiability and containment problems at the level of SHACL constraints are defined in [37], and are shown to be reducible to the problem of SHACL satisfiability. An approach that uses Description Logics Reasoning is presented in [26] to compute *shape containment*, that is, containment at the level of shapes, for a restrictive fragment of SHACL, which however allows recursion.

8   Inference Rules and the Schema Expansion

Datasets are often dynamic objects, which are frequently subject to modification. When an RDF graph is modified, its validity w.r.t. a SHACL document might change. If the modifications that a dataset undergoes are completely arbitrary, then it is not possible to make predictions regarding validity, and the dataset might need to be re-validated after each modification. Many types of modifications that can be applied on a dataset, however, are predictable or a result of some reasoning process. In particular, many types of modifications can be represented as *inference rules* $B \rightarrow H$, where a set of facts H, called the *head* are added to a dataset whenever a query B, called the *body*, finds a match on the dataset. Given an RDF graph G, and a set of inference rules R, it is possible to compute graph G', *closure* of G under R, by applying the *chase* algorithm [4]. The chase algorithm, intuitively, consists in repeatedly applying the rules of R on G until convergence. For simplicity, we assume that the chase algorithm is guaranteed to terminate for the inference rules considered.

Assuming graph G is valid w.r.t. to a SHACL document M, the approach presented in [38], called *schema expansion*, allows us to predict whether the graph closure G' will still be valid w.r.t. M without having to validate G' against M. In particular, given a SHACL document M and a set of inference rules R, the schema expansion process computes the "maximal sub-document" of M which will still validate G after the rule applications. That is, the schema expansion is a SHACL document M', called *schema consequence*, such that (1) $M \subseteq M'$ (i.e., M' is a subset of the restrictions of M); (2) validity is preserved after closure, that is, for any graph G valid w.r.t. M, its closure G' under R is valid w.r.t. M'; and (3) M' is "minimally-containing", i.e., there is no document M'' that satisfies conditions (1) and (2) and such that that $M'' \subset M'$. If a schema consequence M' of a SHACL document M under inference rules R is semantically equivalent to M, then any graph G, valid w.r.t. M is guaranteed to remain valid w.r.t. M after computing its closure under R. In other words, this means that the application of rules R cannot "invalidate" graphs valid w.r.t. document M.

Consider, for example, the following graph G_1, which describes :Eve, a manager of the company in the IT department, and one of her subordinates :Fiona.

```
:Eve a :Manager ;
    :hasDepartment "IT" .
:Fiona a :Employee ;
    :hasManager :Eve .
```

This graph is valid w.r.t. the following SHACL document M_1, which states that each employee must have a manager, and each manager must have a department.

```
:SubordinateS a :PropertyShape ;
    sh:targetClass :Employee ;
    sh:path :hasManager ;
    sh:minCount 1 .

:ManagerS a sh:PropertyShape ;
    sh:targetClass :Manager ;
    sh:path :hasDepartment ;
    sh:minCount 1 .
```

Consider now the set of inference rules $R_1 = \{r_1, r_2\}$, where rules r_1 and r_2 are defined as follows. For simplicity, we represent both the head and the body of rules as SPARQL graph patterns, which are interpreted as SPARQL CONSTRUCT queries where the WHERE and CONSTRUCT clauses are the body and the head, respectively. Rule r_1 states that every manager can be inferred to be an employee, and r_2 states that everyone can be inferred to be in the same department as their manager.

$r_1 = \{\text{?x rdf:type :Manager}\} \rightarrow \{\text{?x rdf:type :Employee}\}$
$r_2 = \{\text{?x :hasManager ?y . ?y :hasDepartment ?z}\} \rightarrow \{\text{?x :hasDepartment ?z}\}$

The closure of graph G_1 under rules R_1 is the following graph G_2.

```
:Eve a :Manager ;
    a :Employee ;
    :hasDepartment "IT" .
:Fiona a :Employee ;
    :hasManager :Eve ;
    :hasDepartment "IT" .
```

Notice that graph G_2 is not valid w.r.t. M_1, since :Eve violates :SubordinateS, but it is valid w.r.t. another document M_2 which only contains shape :ManagerS. In fact, M_2 is a schema consequence of M_1 and R_1. Therefore, we know that the closure under R_1 of any graph valid w.r.t. M_1 will validate shape :ManagerS, but it might not validate :SubordinateS.

Two approaches to compute the schema expansion are presented in [38], for datalog [9] inference rules without negation. The first based on the concept of *critical instance* [30], and the second an optimisation of the first. These approaches are only defined on a fragment of SHACL that, although restricted, is sufficient to express common constraints for RDF validation, such as the Data

Quality Test Patterns TYPEDEP, TYPRODEP, PVT, RDFS-DOMAIN and RDFS-RANGE in the categorisation by Kontokostas et al. [24]. Intuitively, the difficulty in computing a schema expansion lies in having to consider all possible graphs that are valid w.r.t. a SHACL document, and their interactions with arbitrarily complex inference rules.

9   Applications, Tools and Implementations

Over a few years since reaching its status as a W3C recommendation, the level of maturity and adoption of the SHACL technology has been steadily increasing. In this section we review existing SHACL implementations, tools designed to facilitate the creation and management of SHACL documents, and documented usages of SHACL in practical applications.

9.1   Tools for SHACL Validation

The availability of mature tools is often a crucial requirement for the widespread adoption of a technology. To date, SHACL validation has been integrated in a number of mainstream tools and triplestores.[2] An example of this is RDF4J,[3] a Java framework for managing RDF data, which now includes an engine for SHACL validation. The RDF4J framework is integrated in a number of projects, most notably the GraphDB[4] triplestore. Other SHACL-enabled databases include AllegroGraph[5] by Franz Inc, Apache Jena[6] by Apache, and Stardog[7] by Stardog Union Inc. A benchmark for the comparison of different SHACL implementation was proposed in [41], along with results for four different databases. A SHACL implementation is also available for Python through the pySHACL[8] library.

One of the first tools to enable the validation of recursive SHACL graphs was SHACL2SPARQL [11]. Another tool, Trav-SHACL [18], implements a SHACL engine designed to optimise the evaluation of SHACL core constraint components expressible in fragments of the $\mathcal{L}$ language [13]. On these fragments of SHACL, Trav-SHACL was shown to achieve significantly faster validation times compared to the SHACL2SPARQL tool.

9.2   Tools for Generating SHACL Documents

While efficient tools to perform graph validation are undoubtedly essential to the widespread adoption of SHACL, it is also important to devise practical ways to

---

[2] https://w3c.github.io/data-shapes/data-shapes-test-suite/ accessed on 18/6/21.

[3] https://rdf4j.org/ accessed on 18/6/21.

[4] https://graphdb.ontotext.com/ accessed on 18/6/21.

[5] https://allegrograph.com/ accessed on 18/6/21.

[6] https://jena.apache.org/ accessed on 18/6/21.

[7] https://www.stardog.com/ accessed on 18/6/21.

[8] https://pypi.org/project/pyshacl/ accessed on 18/6/21.

generate suitable SHACL documents, without which validation would not be possible. On the one hand, SHACL documents can be manually created by experts. Tools to support this manual process can facilitate this, especially when integrated with already established software. An example of this is SHACL4P [17], a plugin for the Protégé ontology editor [31] which includes an editor to create SHACL documents, and a validator that allows users to test the document by validating an ontology with it, and then visualising any constraint violations. Shape Designer [6] is another tool to create SHACL documents that combines a graphical editor, and additional algorithms to create constraints semi-automatically by analysing the data graph. The benefits of different types of visualisations as an aid to the creating and editing of constraints for RDF graphs was studied in [28]. Existing work also investigated the possibility of generating SHACL documents from natural language text [43].

A number of approaches have been designed to automate the creation of SHACL documents. The SHACLearner [33] approach, generates SHACL documents by learning a kind of rules called Inverse Open Path (IOP) Rules from the graph data provided. IOP rules are strongly related to SCL and therefore SHACL. An IOP rule essentially follows the same structure as an SCL constraint axiom, both syntactically and semantically, with the only exception that the iff operator is replaced by a rightward implication. Another approach to automate the creation of SHACL documents is the Astrea-KG Mappings [10]. These mappings consist of a set of manually created mappings from OWL [49] to SHACL, that can be used to automatically generate SHACL documents from OWL ontologies. As described in [34], SHACL documents can also be generated from the axioms defined by ontology design patterns [19]. The approach from [44] generates SHACL documents for the purpose of quality assessment, using the ontology design patterns and data statistics created by the ABSTAT [45] tool as an input. Another similar approach, presented in [8], allows the automatic extraction of SHACL constraints from a SPARQL endpoint, and was tested on the dataset of Europeana[9].

Notably, SHACL documents can also be seen as describing a desirable "schema" for graph data. As such, they can be used as a template to generate new RDF data. An example of this is the Schímatos [51] tool, which generates forms for RDF graph editing based on SHACL documents, in order to simplify the graph editing task, and minimize the chance of error.

9.3 Adoption of SHACL

An analysis of existing use cases of SHACL can be useful to gain insights on how this technology is used in practice, and on in its level of adoption. In a recent review, 13 existing projects using SHACL have been reviewed, and the most common constraints observed were cardinality, class, datatype and disjunction [29]. Several works investigate the use of SHACL to verify compliance of a dataset w.r.t. certain policies, such as GDPR requirements [2,35]. Other applications of SHACL include type checking program code [27] and detecting metadata errors

---

[9] https://www.europeana.eu/en accessed on 18/6/21.

in clinical studies [22]. SHACL is also used by the European Commission to facilitate data sharing, for example by validating metadata about public services against the recommended vocabularies [46]. Notably, several approaches define translations into SHACL from other technologies, such as ontologies and other schema and constraint languages [16,21,25,32,40,47]. These results show that SHACL tools, and in particular validators, can benefit areas where technologies other than SHACL are already established.

10   Conclusion

Within this review we examined SHACL, a constraint language that can be used to validate RDF graphs. These constraints can be used to describe the properties of a graph, to detect possible errors in the data or provide data quality assurances. In this review we first presented the main concepts of the SHACL specification, such as the concept of *shapes*, and their two main components, *targets* and *constraints*. We discussed the primary way to perform SHACL validation, using SPARQL queries, and how the semantics of validation can be abstracted with the concept of *assignments*.

While the SHACL specification describes how validation should be performed, its semantics is left implicit and not formally defined. We have extensively discussed studies that address this problem. In particular, we reviewed a complete formalisation of SHACL into a fragment of first order logic called SCL. This formalisation lays bare several properties of SHACL, and provides decidability and complexity results for several SHACL-related decision problems. Another important line of work focuses on defining potential extensions of SHACL semantics that can be used in the face of *recursion*. The SHACL specification, in fact, allows constraints to be recursively defined, but it does not define its semantics. We also presented existing work studying the interaction of SHACL with inference rules. Datasets are often dynamic objects, and several questions arise when considering the effects of this dynamism on the constraints imposed over them.

From the point of view of maturity and level of adoption of the SHACL technology, we reviewed several implementations of SHACL validators, which are now integrated in many mainstream RDF databases, and several tools designed to facilitate the creation and management of SHACL documents. Several approaches, in particular, provide automated or semi-automated ways of generating suitable SHACL documents from a diverse range of sources, such as graph data, ontologies, or natural language texts. Existing efforts in mapping other constraint/validation languages into SHACL is also worth noting, as it suggests that the usefulness of SHACL could be extended to support other existing technologies. While the true extent of SHACL adoption is hard to establish, since not all usages of SHACL are publicly documented, we found evidence of its usage in several areas, such as to facilitate data sharing, to validate dataset against policies, and to detect errors in datasets.

Despite the wealth of work on this topic, SHACL is still a recent specification, and a number of important directions for future work still exist. For example, there are opportunities to optimise SHACL validators for particular type of constraints, or for particular scenarios, like for highly dynamic databases. More studies are needed to properly assess the usage of SHACL in practical applications, and what types of constraints are more commonly used and how. While the full semantics of SHACL has been formally defined, more work is needed to formally establish its relation with other constraint languages. It is also important to notice that most of the pieces of work reviewed in this article limit their scope to ad-hoc subsets of the SHACL specification. In addition to the custom requirements of each application, this is commonly done in order to avoid excessively complex language components. At the same time, it is often difficult to understand what these subsets exactly are as they are not always explicitly defined. Therefore, there might be scope to define reusable fragments of SHACL, that could fill the role of lightweight but expressive alternatives to the full language, similarly to how OWL fragments are defined. It might also be beneficial, for similar reasons, to converge towards a single standard or "preferred" semantics for SHACL recursion, which could be defined in a future version of the specification.

References

1. Amarilli, A., Benedikt, M., Bourhis, P., Vanden Boom, M.: Query answering with transitive and linear-ordered data. In: IJCAI 2016, pp. 893–899 (2016)
2. Al Bassit, A., Krasnashchok, K., Skhiri, S., Mustapha, M.: Automated Compliance Checking with SHACL (2020)
3. Andresel, M., Corman, J., Ortiz, M., Reutter, J.L., Savkovic, O., Simkus, M.: Stable model semantics for recursive SHACL. In: Proceedings of The Web Conference 2020, pp. 1570–1580. WWW 2020 (2020)
4. Benedikt, M., et al.: Benchmarking the chase. In: Proceedings of the 36th ACM SIGMOD-SIGACT-SIGAI Symposium on Principles of Database Systems, pp. 37–52. ACM (2017)
5. Berger, R.: The undecidability of the domino problem. MAMS **66**, 1–72 (1966)
6. Boneva, I., Dusart, J., Fernández Alvarez, D., Gayo, J.E.L.: Shape designer for ShEx and SHACL constraints. In: ISWC 2019–18th International Semantic Web Conference, Poster, October 2019
7. Brickley, D., Guha, R.: RDF schema 1.1. W3C recommendation, W3C, February 2014. https://www.w3.org/TR/2014/REC-rdf-schema-20140225/
8. Čerāns, K., Ovčinnikova, J., Bojārs, U., Grasmanis, M., Lāce, L., Romāne, A.: Schema-Backed Visual Queries over Europeana and other Linked Data Resources. In: ESWC2021 Poster and Demo Track (2021)
9. Ceri, S., Gottlob, G., Tanca, L.: What you always wanted to know about datalog (and never dared to ask). IEEE Trans. Knowl. Data Eng. **1**(1), 146–166 (1989)
10. Cimmino, A., Fernández-Izquierdo, A., García-Castro, R.: Astrea: automatic generation of SHACL shapes from ontologies. In: The Semantic Web, pp. 497–513. Springer International Publishing, Cham (2020)
11. Corman, J., Florenzano, F., Reutter, J.L., Savkovic, O.: SHACL2SPARQL: validating a SPARQL endpoint against recursive SHACL constraints. In: International Semantic Web Conference ISWC Satellite Events, pp. 165–168 (2019)

12. Corman, J., Florenzano, F., Reutter, J.L., Savković, O.: Validating SHACL constraints over a SPARQL endpoint. In: Ghidini, C., et al. (eds.) ISWC 2019. LNCS, vol. 11778, pp. 145–163. Springer, Cham (2019). https://doi.org/10.1007/978-3-030-30793-6_9

13. Corman, J., Reutter, J.L., Savković, O.: Semantics and Validation of Recursive SHACL. In: Vrandečić, D., et al. (eds.) ISWC 2018. LNCS, vol. 11136, pp. 318–336. Springer, Cham (2018). https://doi.org/10.1007/978-3-030-00671-6_19

14. Cyganiak, R., Wood, D., Markus Lanthaler, G.: RDF 1.1 concepts and abstract syntax. W3C Recommendation, W3C (2014). http://www.w3.org/TR/2014/REC-rdf11-concepts-20140225/

15. Danielski, D., Kieronski, E.: Finite Satisfiability of Unary Negation Fragment with Transitivity. In: MFCS 2019, pp. 17:1–15. LIPIcs 138, Leibniz-Zentrum fuer Informatik (2019)

16. Di Ciccio, C., Ekaputra, F.J., Cecconi, A., Ekelhart, A., Kiesling, E.: Finding Noncompliances with Declarative Process Constraints Through Semantic Technologies. In: Cappiello, C., Ruiz, M. (eds.) Information Systems Engineering in Responsible Information Systems, pp. 60–74 (2019)

17. Ekaputra, F.J., Lin, X.: SHACL4P: SHACL constraints validation within Protégé ontology editor. In: 2016 International Conference on Data and Software Engineering (ICoDSE), pp. 1–6 (2016)

18. Figuera, M., Rohde, P.D., Vidal, M.E.: Trav-SHACL: efficiently validating networks of SHACL constraints. In: Proceedings of the Web Conference 2021, pp. 3337–3348. WWW 2021, Association for Computing Machinery, New York, NY, USA (2021)

19. Gangemi, A., Presutti, V.: Ontology Design Patterns, pp. 221–243. Springer, Berlin Heidelberg, Berlin, Heidelberg (2009)

20. Jung, J., Lutz, C., Martel, M., Schneider, T.: Querying the Unary Negation Fragment with Regular Path Expressions. In: ICDT 2018, pp. 15:1–18. OpenProceedings.org (2018)

21. K Soman, R.: Modelling construction scheduling constraints using shapes constraint language (SHACL). In: 2019 European Conference on Computing in Construction, pp. 351–358. University College Dublin (2019)

22. Keuchel, D., Spicher, N.: Automatic detection of metadata errors in a registry of clinical studies using shapes constraint language (SHACL) graphs. Stud. Health Technol. Inform. **281**, 372–376 (2021)

23. Knublauch, H., Kontokostas, D.: Shapes Constraint Language (SHACL). W3C Recommendation, W3C (2017). https://www.w3.org/TR/shacl/

24. Kontokostas, D., et al.: Test-driven evaluation of linked data quality. In: Proceedings of the 23rd International Conference on World Wide Web, pp. 747–758. WWW 2014. ACM (2014)

25. Larhrib, M., Escribano, M., Cerrada, C., Escribano, J.J.: Converting OCL and CGMES rules to SHACL in smart grids. IEEE Access **8**, 177255–177266 (2020)

26. Leinberger, M., Seifer, P., Rienstra, T., Lämmel, R., Staab, S.: Deciding SHACL shape containment through description logics reasoning. In: Pan, J.Z., et al. (eds.) ISWC 2020. LNCS, vol. 12506, pp. 366–383. Springer, Cham (2020). https://doi.org/10.1007/978-3-030-62419-4_21

27. Leinberger, M., Seifer, P., Schon, C., Lämmel, R., Staab, S.: Type checking program code using SHACL. In: Ghidini, C., et al. (eds.) ISWC 2019. LNCS, vol. 11778, pp. 399–417. Springer, Cham (2019). https://doi.org/10.1007/978-3-030-30793-6_23

28. Lieber, S., et al.: Visual Notations for Viewing and Editing RDF Constraints with UnSHACLed. Semantic Web (2021). (under review)

29. Lieber, S., Dimou, A., Verborgh, R.: Statistics about data shape use in RDF data. In: ISWC (Demos/Industry) (2020)
30. Marnette, B.: Generalized schema-mappings: from termination to tractability. In: Proceedings of the Twenty-eighth ACM SIGMOD-SIGACT-SIGART Symposium on Principles of Database Systems, pp. 13–22. ACM (2009)
31. Musen, M.A.: Protégé Team: The Protégé Project: A Look Back and a Look Forward. AI Matters **1**(4), 4–12 (2015)
32. Nenadić, K.R., Gavrić, M.M., Durdević, V.I.: Validation of CIM datasets using SHACL. In: 2017 25th Telecommunication Forum (TELFOR), pp. 1–4 (2017)
33. Omran, P.G., Taylor, K., Mendez, S.R., Haller, A.: Learning SHACL Shapes from Knowledge Graphs. Semantic Web (2021). (under review)
34. Pandit, H.J., O'Sullivan, D., Lewis, D.: Using ontology design patterns to define SHACL shapes. In: 9th Workshop on Ontology Design and Patterns (WOP2018), International Semantic Web Conference (ISWC), pp. 67–71 (2018)
35. Pandit, H.J., O'Sullivan, D., Lewis, D.: Test-driven approach towards GDPR compliance. In: Acosta, M., et al. (eds.) SEMANTiCS 2019. LNCS, vol. 11702, pp. 19–33. Springer, Cham (2019). https://doi.org/10.1007/978-3-030-33220-4_2
36. Pareti, P., Konstantinidis, G., Mogavero, F.: Satisfiability and Containment of Recursive SHACL (2021). arXiv preprint 2108.13063
37. Pareti, P., Konstantinidis, G., Mogavero, F., Norman, T.J.: SHACL satisfiability and containment. In: Pan, J.Z., et al. (eds.) ISWC 2020. LNCS, vol. 12506, pp. 474–493. Springer, Cham (2020). https://doi.org/10.1007/978-3-030-62419-4_27
38. Pareti, P., Konstantinidis, G., Norman, T.J., Şensoy, M.: SHACL constraints with inference rules. In: Ghidini, C., et al. (eds.) ISWC 2019. LNCS, vol. 11778, pp. 539–557. Springer, Cham (2019). https://doi.org/10.1007/978-3-030-30793-6_31
39. Robinson, R.: Undecidability and nonperiodicity for tilings of the plane. IM **12**, 177–209 (1971)
40. Savković, O., Kharlamov, E., Lamparter, S.: Validation of SHACL constraints over KGs with OWL 2 QL ontologies via rewriting. In: Hitzler, P., et al. (eds.) ESWC 2019. LNCS, vol. 11503, pp. 314–329. Springer, Cham (2019). https://doi.org/10.1007/978-3-030-21348-0_21
41. Schaffenrath, R., et al.: Benchmark for performance evaluation of SHACL implementations in graph databases. In: Gutiérrez-Basulto, V., Kliegr, T., Soylu, A., Giese, M., Roman, D. (eds.) RuleML+RR 2020. LNCS, vol. 12173, pp. 82–96. Springer, Cham (2020). https://doi.org/10.1007/978-3-030-57977-7_6
42. Seaborne, A., Harris, S.: SPARQL 1.1 query language. W3C recommendation, W3C, March 2013. https://www.w3.org/TR/2013/REC-sparql11-query-20130321/
43. Šenkýř, D.: SHACL shapes generation from textual documents. In: Pergl, R., Babkin, E., Lock, R., Malyzhenkov, P., Merunka, V. (eds.) EOMAS 2019. LNBIP, vol. 366, pp. 121–130. Springer, Cham (2019). https://doi.org/10.1007/978-3-030-35646-0_9
44. Spahiu, B., Maurino, A., Palmonari, M.: Towards improving the quality of knowledge graphs with data-driven ontology patterns and SHACL. In: International Semantic Web Conference ISWC Sattelite Events, pp. 103–117 (2018)
45. Spahiu, B., Porrini, R., Palmonari, M., Rula, A., Maurino, A.: ABSTAT: ontology-driven linked data summaries with pattern minimalization. In: Sack, H., et al. (eds.) ESWC 2016. LNCS, vol. 9989, pp. 381–395. Springer, Cham (2016). https://doi.org/10.1007/978-3-319-47602-5_51
46. Stani, E.: Design reusable SHACL shapes and implement a linked data validation pipeline. Code4Lib. J. **45**, 1–12 (2019)

47. Stolk, S., McGlinn, K.: Validation of IfcOWL datasets using SHACL. In: Proceedings of the 8th Linked Data in Architecture and Construction Workshop, pp. 91–104 (2020)

48. Thornton, K., et al.: Using shape expressions (ShEx) to share RDF data models and to guide curation with rigorous validation. In: Hitzler, P., et al. (eds.) ESWC 2019. LNCS, vol. 11503, pp. 606–620. Springer, Cham (2019). https://doi.org/10.1007/978-3-030-21348-0_39

49. W3C OWL Working Group: OWL 2 Web Ontology Language Document Overview (Second Edition). W3C recommendation, W3C (Dec 2012). https://www.w3.org/TR/2012/REC-owl2-overview-20121211/

50. Wang, H.: Proving theorems by pattern recognition II. BSTJ **40**, 1–41 (1961)

51. Wright, J., Rodríguez Méndez, S.J., Haller, A., Taylor, K., Omran, P.G.: *Schímatos*: a SHACL-based web-form generator for knowledge graph editing. In: Pan, J.Z., et al. (eds.) ISWC 2020. LNCS, vol. 12507, pp. 65–80. Springer, Cham (2020). https://doi.org/10.1007/978-3-030-62466-8_5

Score-Based Explanations in Data Management and Machine Learning: An Answer-Set Programming Approach to Counterfactual Analysis

Leopoldo Bertossi[1,2(✉)]

[1] Faculty of Engineering and Sciences, Universidad Adolfo Ibáñez, Santiago, Chile
leopoldo.bertossi@uai.cl
[2] Millennium Inst. for Foundational Research on Data (IMFD), Santiago, Chile

Abstract. We describe some recent approaches to score-based explanations for query answers in databases and outcomes from classification models in machine learning. The focus is on work done by the author and collaborators. Special emphasis is placed on declarative approaches based on answer-set programming to the use of counterfactual reasoning for score specification and computation. Several examples that illustrate the flexibility of these methods are shown.

1 Introduction

In data management and machine learning one wants *explanations* for certain results. For example, for query results from databases, and for outcomes of classification models in machine learning (ML). Explanations, that may come in different forms, have been the subject of philosophical enquires for a long time, but, closer to our discipline, they appear under different forms in model-based diagnosis and in causality as developed in artificial intelligence.

In the last few years, explanations that are based on *numerical scores* assigned to elements of a model that may contribute to an outcome have become popular. These scores attempt to capture the degree of contribution of those components to an outcome, e.g. answering questions like these: What is the contribution of this tuple to the answer to this query? What is the contribution of this feature value of an entity to the displayed classification of the latter?

For an example, consider a financial institution that uses a learned classifier, $\mathcal{C}$, e.g. a decision tree, to determine if clients should be granted loans or not, returning labels 0 or 1, resp. A particular client, represented as an entity $\mathbf{e}$, applies for a loan, and the classifier returns $\mathcal{C}(\mathbf{e}) = 1$, i.e. the loan is rejected. The client requests an explanation.

A common approach consists in giving scores to the feature values in $\mathbf{e}$, to quantify their relevance in relation to the classification outcome. The higher the score of a feature value, the more explanatory is that value. For example, the fact that the client has value "5" for feature *Age* (in years) could have the highest

© Springer Nature Switzerland AG 2022
M. Šimkus and I. Varzinczak (Eds.): Reasoning Web 2021, LNCS 13100, pp. 145–184, 2022.
https://doi.org/10.1007/978-3-030-95481-9_7

score. That is, the rejection of the loan application is due mostly to the client's very young age.

In the context of *explainable AI* [39], different scores have been proposed in the literature, and some that have a relatively older history have been applied. Among the latter we find the general *responsibility score* as found in *actual causality* [19,26]. For a particular kind of application, one has to define the right causality setting, and then apply the responsibility measure to the participating variables (see [27] for a newer treatment of the subject). In particular, in data management, responsibility has been used to quantify the strength of a tuple as a cause for a query result [5,36]. The Shapley value, as found in *coalition game theory* [45], has been used for the same purpose [30]. Defining the right game function, the Shapley value assigned to a player reflects its contribution to the wealth function, which in databases corresponds to the query result.

In the context of explanations to outcomes from classification models in ML, the Shapley value has been used to assign scores to the feature values taken by an entity that has been classified. With a particular game function, it has taken the form of the Shap score, which has become quite popular and influential [34,35].

Also recently, a *responsibility score*, Resp, has been introduced and investigated for the same purpose in [9]. It is based on the notions of *counterfactual intervention* as appearing in actual causality, and causal responsibility. More specifically, (potential) executions of *counterfactual interventions* on a *structural logico-probabilistic model* [26] are investigated, with the purpose of answering hypothetical questions of the form: *What would happen if we change ...?*.

Counterfactual interventions can be used to define different forms of score-based explanations. This is the case of causal responsibility in databases (c.f. Sect. 12). In *explainable AI*, and more commonly with classification models of ML, counterfactual interventions become hypothetical changes on the entity whose classification is being explained, to detect possible changes in the outcome (c.f. [11, Sect. 8] for a more detailed discussion and references).

Score-based explanations can also be defined in the absence of a model, and with or without *explicit* counterfactual interventions. Actually, explanation scores such as Shap and Resp can be applied with *black-box* models, in that they use, in principle, only the input/output relation that represents the classifier, without having access to the internal components of the model. In this category we could find classifiers based on complex neural networks, or XGBoost [33]. They are opaque enough to be treated as black-box models.

The Shap and Resp scores can also be applied with open-box models, with explicit models. Without having access to the elements of the classification model, the computation of both Shap and Resp is in general intractable, by their sheer definitions, and the possibly large number of counterfactual combinations that have to be considered in the computation. However, for certain classes of classifiers, e.g. decision trees, having access to the mathematical model may make the computation of Shap tractable, as shown in [3,48], where it is also shown that for other classes of explicit models, its computation is still intractable. Something similar applies to Resp [9].

Other explanation scores used in machine learning appeal to the components of the mathematical model behind the classifier. There can be all kinds of explicit models, and some are easier to understand or interpret or use for this purpose. For example, the *FICO score* proposed in [18], for the FICO dataset about loan requests, depends on the internal outputs and displayed coefficients of two nested logistic regression models. Decision trees [38], random forests [12], rule-based classifiers, etc., could be seen as relatively easy to understand and use for providing explanations. In [9], the Shap and Resp scores were experimentally compared with each other, and also with the FICO score.

One can specify in declarative terms the counterfactual versions of tuples in databases and of feature values in entities under classification. On this basis one can analyze diverse alternative counterfactuals, reason about them, and also specify the associated explanation scores. In these notes we do this for responsibility scores in databases and classifications models. More specifically, we use *answer-set programming*, a modern logic-programming paradigm that has become useful in many applications [13,24]. We show examples run with the *DLV* system and its extensions [29]. An important advantage of using declarative specifications resides in the possibility of adding different forms of *domain knowledge* and *semantic constraints*. Doing this with purely procedural approaches would require changing the code accordingly.

The answer-set programs (ASPs) we use are influenced by, and sometimes derived from, *repair programs*. These are ASPs that specify and compute the possible repairs of a database that is inconsistent with respect to a given set of integrity constraints [4]. A useful connection between database repairs and actual causality in databases was established in [5]. Hence, the use of repairs and repair programs.

In this article we survey some of the recent advances on the use and computation of the above mentioned score-based explanations, both for query answering in databases and for classification in ML. This is not intended to be an exhaustive survey of the area. Instead, it is heavily influenced by our latest research. Special emphasis is placed on the use of ASPs (for many more details on this see [11]). Taking advantage of the introduced repair programs, we also show how to specify and compute a numerical measure of inconsistency of database [7]. In this case, this would be a *global* score, in contrast with the *local* scores applied to individual tuples in a database or feature values in an entity. To introduce the concepts and techniques we will use mostly examples, trying to convey the main intuitions and issues.

This paper is structured as follows. In Sect. 2 we provide some background material on databases and answer-set programs. In Sect. 3 we concentrate on causal explanations in databases, the responsibility score, and also the *causal-effect* score [44], as an alternative to the latter. In Sect. 4, we present the causality-repair connection and repair programs for causality and responsibility computation. In Sect. 5, we consider causality in databases at the attribute level, as opposed to the tuple level. In Sect. 6, we introduce causality and responsibility in databases that are subject to integrity constraints. In Sect. 7 we present the global inconsistency measure for a database and the ASPs to compute it. In Sect. 8, we describe the use of the Shapley value to provide explanation scores in databases. In Sect. 9, we describe in general terms score-based explanations for classification results. In Sect. 10 we introduce and study the x-Resp score, a simpler version of the more general Resp score that we introduce in Sect. 12. In Sect. 11 we introduce *counterfactual intervention programs* (CIP), which are ASPs that specify counterfactuals and the x-Resp score. In Sect. 13, and for completeness, we briefly present the Shap score. We end in Sect. 14 with some final conclusions.

2 Background

2.1 Basics of Relational Databases

A relational schema $\mathcal{R}$ contains a domain of constants, $\mathcal{C}$, and a set of predicates of finite arities, $\mathcal{P}$. $\mathcal{R}$ gives rise to a language $\mathfrak{L}(\mathcal{R})$ of first-order (FO) predicate logic with built-in equality, $=$. Variables are usually denoted with $x, y, z, ...$; and finite sequences thereof with $\bar{x}, ...$; and constants with $a, b, c, ...$, etc. An *atom* is of the form $P(t_1, \ldots, t_n)$, with n-ary $P \in \mathcal{P}$ and $t_1, \ldots, t_n$ *terms*, i.e. constants, or variables. An atom is *ground* (a.k.a. a tuple) if it contains no variables. A database (instance), D, for $\mathcal{R}$ is a finite set of ground atoms; and it serves as an interpretation structure for $\mathfrak{L}(\mathcal{R})$.

A *conjunctive query* (CQ) is a FO formula, $\mathcal{Q}(\bar{x})$, of the form $\exists \bar{y} \, (P_1(\bar{x}_1) \land \cdots \land P_m(\bar{x}_m))$, with $P_i \in \mathcal{P}$, and (distinct) free variables $\bar{x} := (\bigcup \bar{x}_i) \smallsetminus \bar{y}$. If $\mathcal{Q}$ has n (free) variables, $\bar{c} \in \mathcal{C}^n$ is an *answer* to $\mathcal{Q}$ from D if $D \models \mathcal{Q}[\bar{c}]$, i.e. $Q[\bar{c}]$ is true in D when the variables in $\bar{x}$ are componentwise replaced by the values in $\bar{c}$. $\mathcal{Q}(D)$ denotes the set of answers to $\mathcal{Q}$ from D. $\mathcal{Q}$ is a *Boolean conjunctive query* (BCQ) when $\bar{x}$ is empty; and when *true* in D, $\mathcal{Q}(D) := \{true\}$. Otherwise, it is *false*, and $\mathcal{Q}(D) := \emptyset$. Sometimes CQs are written in Datalog notation as follows: $\mathcal{Q}(\bar{x}) \leftarrow P_1(\bar{x}_1), \ldots, P_m(\bar{x}_m)$.

We consider as integrity constraints (ICs), i.e. sentences of $\mathfrak{L}(\mathcal{R})$: (a) *denial constraints* (DCs), i.e. of the form $\kappa : \neg \exists \bar{x}(P_1(\bar{x}_1) \land \cdots \land P_m(\bar{x}_m))$, where $P_i \in \mathcal{P}$, and $\bar{x} = \bigcup \bar{x}_i$; and (b) *functional dependencies* (FDs), i.e. of the form $\varphi : \neg \exists \bar{x}(P(\bar{v}, \bar{y}_1, z_1) \land P(\bar{v}, \bar{y}_2, z_2) \land z_1 \neq z_2)$.[1] Here, $\bar{x} = \bar{y}_1 \cup \bar{y}_2 \cup \bar{v} \cup \{z_1, z_2\}$, and $z_1 \neq z_2$ is an abbreviation for $\neg z_1 = z_2$. A *key constraint* (KC) is a conjunction of FDs: $\bigwedge_{j=1}^{k} \neg \exists \bar{x}(P(\bar{v}, \bar{y}_1) \land P(\bar{v}, \bar{y}_2) \land y_1^j \neq y_2^j)$, with $k = |\bar{y}_1| = |\bar{y}_2|$, and generically y^j stands for the jth variable in $\bar{y}$. For example, $\forall x \forall y \forall z (Emp(x, y) \land Emp(x, z) \rightarrow y = z)$, is an FD (and also a KC) that could say that an employee (x) can have at most one salary. This FD is usually written as $EmpName \rightarrow EmpSalary$. In the following, we will include FDs and key constraints among the DCs.

We will also consider *inclusion dependencies* (INDs), which are constraints of the form $\forall \bar{x} \exists \bar{y}(P_1(\bar{x}) \rightarrow P_2(\bar{x}', \bar{y}))$, where $P_1, P_2 \in \mathcal{P}$, and $\bar{x}' \subseteq \bar{x}$.

---

[1] The variables in $\bar{v}$ do not have to go first in the atomic formulas; what matters is keeping the correspondences between the variables in those formulas.

If an instance D does not satisfy the set Σ of ICs associated to the schema, we say that D is *inconsistent*, which is denoted with $D \not\models \Sigma$.

2.2   Basics of Answer-Set Programming

We will give now a brief review of the basics of *answer-set programs* (ASPs). As customary, when we talk about ASPs, we refer to *disjunctive Datalog programs with weak negation and stable model semantics* [23,24]. For this reason we will, for a given program, use the terms "stable model" (or simply, "model") and "answer-set" interchangeably. An answer-set program Π consists of a finite number of rules of the form

$$A_1 \vee \ldots \vee A_n \leftarrow P_1, \ldots, P_m, not\ N_1, \ldots, not\ N_k, \tag{1}$$

where $0 \leq n, m, k$, and A_i, P_j, N_s are (positive) atoms, i.e. of the form $Q(\bar{t})$, where Q is a predicate of a fixed arity, say, ℓ, and $\bar{t}$ is a sequence of length ℓ of variables or constants. In rule (6), $A_1, \ldots, not\ N_k$ are called *literals*, with A_1 *positive*, and $not\ N_k$, *negative*. All the variables in the A_i, N_s appear among those in the P_j. The left-hand side of a rule is called the *head*, and the right-hand side, the *body*. A rule can be seen as a (partial) definition of the predicates in the head (there may be other rules with the same predicates in the head).

The constants in program Π form the (finite) Herbrand universe H of the program. The ground version of program Π, $gr(\Pi)$, is obtained by instantiating the variables in Π in all possible ways using values from H. The Herbrand base, HB, of Π contains all the atoms obtained as instantiations of predicates in Π with constants in H.

A subset M of HB is a model of Π if it satisfies $gr(\Pi)$, i.e.: For every ground rule $A_1 \vee \ldots \vee A_n \leftarrow P_1, \ldots, P_m, not\ N_1, \ldots, not\ N_k$ of $gr(\Pi)$, if $\{P_1, \ldots, P_m\} \subseteq M$ and $\{N_1, \ldots, N_k\} \cap M = \emptyset$, then $\{A_1, \ldots, A_n\} \cap M \neq \emptyset$. M is a minimal model of Π if it is a model of Π, and Π has no model that is properly contained in M. $MM(\Pi)$ denotes the class of minimal models of Π. Now, for $S \subseteq HB(\Pi)$, transform $gr(\Pi)$ into a new, positive program $gr(\Pi)^S$ (i.e. without *not*), as follows: Delete every rule $A_1 \vee \ldots \vee A_n \leftarrow P_1, \ldots, P_m, not\ N_1, \ldots, not\ N_k$ for which $\{N_1, \ldots, N_k\} \cap S \neq \emptyset$. Next, transform each remaining rule $A_1 \vee \ldots \vee A_n \leftarrow P_1, \ldots, P_m, not\ N_1, \ldots, not\ N_k$ into $A_1 \vee \ldots \vee A_n \leftarrow P_1, \ldots, P_m$. Now, S is a *stable model* of Π if $S \in MM(gr(\Pi)^S)$. Every stable model of Π is also a minimal model of Π. Stable models are also commonly called *answer sets*, and so are we going to do most of the time.

A program is *unstratified* if there is a cyclic, recursive definition of a predicate that involves negation. For example, the program consisting of the rules $a \vee b \leftarrow c, not\ d;$ $d \leftarrow e,$ and $e \leftarrow b$ is unstratified, because there is a negation in the mutually recursive definitions of b and e. The program in Example 1 below is not unstratified, i.e. it is *stratified*. A good property of stratified programs is that the models can be upwardly computed following *strata* (layers) starting from the *facts*, that is from the ground instantiations of rules with empty bodies (in which case the arrow is usually omitted). We refer the reader to [24] for more details.

Query answering under the ASPs comes in two forms. Under the *brave seman-tics*, a query posed to the program obtains as answers those that hold in *some* model of the program. However, under the *skeptical* (or *cautious*) semantics, only the answers that simultaneously hold in *all* the models are returned. Both are useful depending on the application at hand.

Example 1. Consider the following program Π that is already ground.

$$a \vee b \leftarrow c$$
$$d \leftarrow b$$
$$a \vee b \leftarrow e, \; notf$$
$$e \leftarrow$$

The program has two stable mod-els: $S_1 = \{e, a\}$ and $S_2 = \{e, b, d\}$.
Each of them expresses that the atoms in it are true, and any other atom that does not belong to it, is false.

These models are incomparable under set inclusion, and are minimal models in that any proper subset of any of them is not a model of the program (i.e. does not satisfy the program). □

3   Causal Explanations in Databases

In data management we need to understand and compute *why* certain results are obtained or not, e.g. query answers, violations of semantic conditions, etc.; and we expect a database system to provide *explanations*.

3.1   Causal Responsibility

Here, we will consider *causality-based explanations* [36,37], which we will illus-trate by means of an example.

Example 2. Consider the database D, and the Boolean conjunctive query (BCQ)

R	A	B
	a	b
	c	d
	b	b

S	C
	a
	c
	b

$$Q: \; \exists x \exists y (S(x) \wedge R(x, y) \wedge S(y)). \quad (2)$$

It holds: $D \models Q$, i.e. the query is true in D.

We ask about the causes for Q to be true: A tuple $\tau \in D$ is *counterfactual cause* for Q (being true in D) if $D \models Q$ and $D \smallsetminus \{\tau\} \not\models Q$. In this example, $S(b)$ is a counterfactual cause for Q: If $S(b)$ is removed from D, Q is no longer true.

Removing a single tuple may not be enough to invalidate the query. Accord-ingly, a tuple $\tau \in D$ is an *actual cause* for Q if there is a *contingency set* $\Gamma \subseteq D$, such that τ is a counterfactual cause for Q in $D \smallsetminus \Gamma$. In this example, $R(a, b)$ is an actual cause for Q with contingency set $\{R(b, b)\}$: If $R(a, b)$ is removed from D, Q is still true, but further removing $R(b, b)$ makes Q false. □

Notice that every counterfactual cause is also an actual cause, with empty contingent set. Actual causes that are not counterfactual causes need company to invalidate a query result. Now we ask how strong are tuples as actual causes. To answer this question, we appeal to the *responsibility* of an actual cause τ for $\mathcal{Q}$ [36], defined by:

$$\rho_D(\tau) := \frac{1}{|\Gamma| + 1},$$

where $|\Gamma|$ is the size of a smallest contingency set, Γ, for τ, and 0, otherwise.

Example 3 (Example 2 cont.). The responsibility of $R(a,b)$ is $\frac{1}{2} = \frac{1}{1+1}$ (its several smallest contingency sets have all size 1).

$R(b,b)$ and $S(a)$ are also actual causes with responsibility $\frac{1}{2}$; and $S(b)$ is actual (counterfactual) cause with responsibility $1 = \frac{1}{1+0}$.   □

High responsibility tuples provide more interesting explanations. Causes in this case are tuples that come with their responsibilities as "scores". All tuples can be seen as actual causes, but only those with non-zero responsibility score matter. Causality and responsibility in databases can be extended to the attribute-value level [5,8] (c.f. Sect. 5).

As we will see in Sect. 4.1, there is a connection between database causality and *repairs* of databases w.r.t. integrity constraints (ICs) [4]. There are also connections to *consistency-based diagnosis* and *abductive diagnosis*, that are two forms of *model-based diagnosis* [46]. These connections have led to new complexity and algorithmic results for causality and responsibility [5,6]. Actually, the latter turns out to be intractable (c.f. Sect. 4.1). In [6], causality under ICs was introduced and investigated. This allows to bring semantic and domain knowledge into causality in databases (c.f. Sect. 6).

Model-based diagnosis is an older area of knowledge representation where explanations form the subject of investigation. In general, the diagnosis analysis is performed on a logic-based model, and certain elements of the model are identified as explanations. Causality-based explanations are somehow more recent. In this case, still a model is used, which is, in general, a more complex than a database with a query. In the case of databases, actually there is an underlying logical model, the *lineage or provenance* of the query [14,47] that we will illustrate in Sect. 3.2, but it is still a relatively simple model.

3.2 The Causal-Effect Score

Sometimes, as we will see right here below, responsibility does not provide intuitive or expected results, which led to the consideration of an alternative score, the *causal-effect score*. We show the issues and the score by means of an example.

Example 4. Consider the database E that represents the graph below, and the Boolean Datalog query Π that is true in E if there is a path from a to b. Here, $E \cup \Pi \models yes$. Tuples have global tuple identifiers (tids) in the left-most column, which is not essential, but convenient.

E	A	B
t_1	a	b
t_2	a	c
t_3	c	b
t_4	a	d
t_5	d	e
t_6	e	b

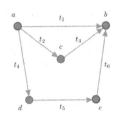

$$yes \leftarrow P(a,b)$$
$$P(x,y) \leftarrow E(x,y)$$
$$P(x,y) \leftarrow P(x,z), E(z,y)$$

All tuples are actual causes since every tuple appears in a path from a to b. Also, all the tuples have the same causal responsibility, $\frac{1}{3}$, which may be counterintuitive, considering that t_1 provides a direct path from a to b.   □

In [44], the notion *causal effect* was introduced. It is based on three main ideas, namely, the transformation, for auxiliary purposes, of the database into a probabilistic database, the expected value of a query, and interventions on the lineage of the query. The lineage of a query represents, by means of a propositional formula, all the ways in which the query can be true in terms of the potential database tuples, and their combinations. Here, "potential" refers to tuples that can be built with the database predicates and the database (finite) domain. These tuples may belong to the database at hand or not. For a given database, D, some of those atoms become true, and others false, which leads to the instantiation of the lineage (formula) o D. This is all shown in the next example.

Example 5. Consider the database D below, and a BCQ.

R	A	B
	a	b
	a	c
	c	b

S	C
	b
	c

$\mathcal{Q}: \exists x \exists y (R(x,y) \wedge S(y))$, which is true in D.

For the database D in our example, the lineage of the query instantiated on D is given by the propositional formula:

$$\Phi_{\mathcal{Q}}(D) = (X_{R(a,b)} \wedge X_{S(b)}) \vee (X_{R(a,c)} \wedge X_{S(c)}) \vee (X_{R(c,b)} \wedge X_{S(b)}), \quad (3)$$

where X_τ is a propositional variable that is true iff $\tau \in D$. Here, $\Phi_{\mathcal{Q}}(D)$ takes value 1 in D.

Now, for illustration, we want to quantify the contribution of tuple $S(b)$ to the query answer. For this purpose, we assign, uniformly and independently, probabilities to the tuples in D, obtaining a *probabilistic database* D^p [47]. Potential tuples outside D get probability 0.

R^p	A	B	prob
	a	b	$\frac{1}{2}$
	a	c	$\frac{1}{2}$
	c	b	$\frac{1}{2}$

S^p	C	prob
	b	$\frac{1}{2}$
	c	$\frac{1}{2}$

The X_τ's become independent, identically distributed Boolean random variables; and $\mathcal{Q}$ becomes a Boolean random variable. Accordingly, we can ask about the probability that $\mathcal{Q}$ takes the truth value 1 (or 0) when an *intervention* is performed on D.

Interventions are of the form $do(X = x)$, meaning making X take value x, with $x \in \{0,1\}$, in the *structural model*, in this case, the lineage. That is, we ask, for $\{y,x\} \subseteq \{0,1\}$, about the conditional probability $P(\mathcal{Q} = y \mid do(X_\tau = x))$, i.e. conditioned to making X_τ false or true.

For example, with $do(X_{S(b)} = 0)$ and $do(X_{S(b)} = 1)$, the lineage in (3) becomes, resp., and abusing the notation a bit:

$$\Phi_\mathcal{Q}(D|do(X_{S(b)} = 0) := (X_{R(a,c)} \wedge X_{S(c)}).$$
$$\Phi_\mathcal{Q}(D|do(X_{S(b)} = 1) := X_{R(a,b)} \vee (X_{R(a,c)} \wedge X_{S(c)}) \vee X_{R(c,b)}.$$

On the basis of these lineages and D^p, when $X_{S(b)}$ is made false, the probability that the instantiated lineage becomes true in D^p is:

$$P(\mathcal{Q} = 1 \mid do(X_{S(b)} = 0)) = P(X_{R(a,c)} = 1) \times P(X_{S(c)} = 1) = \frac{1}{4}.$$

Similarly, when $X_{S(b)}$ is made true, the probability of the lineage becoming true in D^p is:

$$P(\mathcal{Q} = 1 \mid do(X_{S(b)} = 1)) = P(X_{R(a,b)} \vee (X_{R(a,c)} \wedge X_{S(c)}) \vee X_{R(c,b)} = 1) = \frac{13}{16}.$$

The *causal effect* of a tuple τ is defined by:

$$\mathcal{CE}^{D,\mathcal{Q}}(\tau) := \mathbb{E}(\mathcal{Q} \mid do(X_\tau = 1)) - \mathbb{E}(\mathcal{Q} \mid do(X_\tau = 0)).$$

In particular, using the probabilities computed so far:

$$\mathbb{E}(\mathcal{Q} \mid do(X_{S(b)} = 0)) = P(\mathcal{Q} = 1 \mid do(X_{S(b)} = 0)) = \frac{1}{4},$$
$$\mathbb{E}(\mathcal{Q} \mid do(X_{S(b)} = 1)) = P(\mathcal{Q} = 1 \mid do(X_{S(b)} = 1)) = \frac{13}{16}.$$

Then, the causal effect for the tuple $S(b)$ is: $\mathcal{CE}^{D,\mathcal{Q}}(S(b)) = \frac{13}{16} - \frac{1}{4} = \frac{9}{16} > 0$, showing that the tuple is relevant for the query result, with a relevance score provided by the causal effect, of $\frac{9}{16}$. $\qquad \square$

Example 6 (Example 4 cont.). The Datalog query, here as a union of BCQs, has the lineage: $\Phi_\mathcal{Q}(D) = X_{t_1} \vee (X_{t_2} \wedge X_{t_3}) \vee (X_{t_4} \wedge X_{t_5} \wedge X_{t_6})$. It holds:

$$\mathcal{CE}^{D,\mathcal{Q}}(t_1) = 0.65625,$$
$$\mathcal{CE}^{D,\mathcal{Q}}(t_2) = \mathcal{CE}^{D,\mathcal{Q}}(t_3) = 0.21875,$$
$$\mathcal{CE}^{D,\mathcal{Q}}(t_4) = \mathcal{CE}^{D,\mathcal{Q}}(t_5) = \mathcal{CE}^{D,\mathcal{Q}}(t_6) = 0.09375.$$

The causal effects are different for different tuples, and the scores are much more intuitive than the responsibility scores. $\qquad \square$

The definition of the causal-effect score may look rather *ad hoc* and arbitrary. We will revisit it in Sect. 8, where we will have yet another explanation score in databases; namely one that takes a new approach, measuring the contribution of a database tuple to a query answer through the use of the *Shapley value*, which is firmly established in game theory, and is also used in several other areas [43,45].

The main idea is that *several tuples together*, much like players in a coalition game, are necessary to violate an IC or produce a query result. Some may contribute more than others to the *wealth distribution function* (or simply, game function), which in this case becomes the query result, namely 1 or 0 if the query is Boolean, or a number if the query is an aggregation. The Shapley value of a tuple can be used to assign a score to its contribution. This was done in [30], and will be retaken in Sect. 8. But first things first.

4   Answer-Set Programs for Causality in Databases

In this section we will first establish a useful connection between database repairs and causes as tuples in a database. Then, we use ASPs, taking the form of *repair programs*, to specify and compute database repairs and tuples as causes for query answers. We end this section with a fully developed example using the *DLV* system and its extensions [29].

4.1   The Repair Connection

The notion of *repair* of a relational database was introduced in order to formalize the notion of *consistent query answering* (CQA), as shown in Fig. 1: If a database D is inconsistent in the sense that is does not satisfy a given set of integrity constraints, *ICs*, and a query Q is posed to D (left-hand side of Fig. 1), what are the meaningful, or consistent, answers to Q from D? They are sanctioned as those that hold (are returned as answers) from *all* the *repairs* of D. The repairs of D are consistent instances D' (over the same schema of D), i.e. $D' \models ICs$, and *minimally depart* from D [2,4] (right-hand side of Fig. 1).

Notice that: (a) We have now a *possible-world* semantics for (consistent) query answering; and (b) we may use in principle any reasonable notion of distance

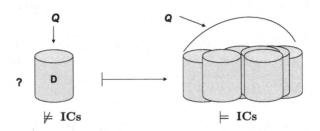

Fig. 1. Database repairs and consistent query answers

between database instances, with each choice defining a particular *repair seman-tics*. In the rest of this section we will illustrate two classes of repairs, which have been used and investigated the most in the literature. Actually, repairs in general have got a life of their own, beyond consistent query answering.

Example 7. Let us consider the following set of *denial constraints* (DCs) and a database D, whose relations (tables) are shown right here below. D is inconsistent, because it violates the DCs: it satisfies the joins that are prohibited by the DCs.

$$\neg\exists x\exists y(P(x) \wedge Q(x,y))$$
$$\neg\exists x\exists y(P(x) \wedge R(x,y))$$

P	A
	a
	e

Q	A	B
	a	b

R	A	C
	a	c

We want to repair the original instance by *deleting tuples* from relations. Notice that, for DCs, insertions of new tuple will not restore consistency. We could change (update) attribute values though, a possibility we will consider in Sect. 5.

Here we have two *subset repairs*, a.k.a. *S-repairs*. They are subset-maximal consistent subinstances of D: $D_1 = \{P(e), Q(a,b), R(a,c)\}$ and $D_2 = \{P(e), P(a)\}$. They are consistent, subinstances of D, and any proper superset of them (still contained in D) is inconsistent. (In general, we will represent database relations as set of tuples).

We also have *cardinality repairs*, a.k.a. *C-repairs*. They are consistent subinstances of D that minimize the *number* of tuples by which they differ from D. That is, they are maximum-cardinality consistent subinstances. In this case, only D_1 is a C-repair. Every C-repair is an S-repair, but not necessarily the other way around (as this example shows). □

Let us now consider a BCQ

$$\mathcal{Q}: \exists\bar{x}(P_1(\bar{x}_1) \wedge \cdots \wedge P_m(\bar{x}_m)), \tag{4}$$

which we assume is true in a database D. It turns out that we can obtain the causes for $\mathcal{Q}$ to be true D, and their contingency sets from database repairs. In order to do this, notice that $\neg\mathcal{Q}$ becomes a DC

$$\kappa(\mathcal{Q}): \neg\exists\bar{x}(P_1(\bar{x}_1) \wedge \cdots \wedge P_m(\bar{x}_m)); \tag{5}$$

and that $\mathcal{Q}$ holds in D iff D is inconsistent w.r.t. $\kappa(\mathcal{Q})$.

It holds that S-repairs are associated to causes with minimal contingency sets, while C-repairs are associated to causes for $\mathcal{Q}$ with minimum contingency sets, and maximum responsibilities [5]. In fact, for a database tuple $\tau \in D$:

(a) τ is actual cause for $\mathcal{Q}$ with subset-minimal contingency set Γ iff $D \smallsetminus (\Gamma \cup \{\tau\})$ is an S-repair (w.r.t. $\kappa(\mathcal{Q})$), in which case, its responsibility is $\frac{1}{1+|\Gamma|}$.

(b) τ is actual cause with minimum-cardinality contingency set Γ iff $D \smallsetminus (\Gamma \cup \{\tau\})$ is C-repair, in which case, τ is a maximum-responsibility actual cause.

Conversely, repairs can be obtained from causes and their contingency sets [5]. These results can be extended to unions of BCQs (UBCQs), or equivalently, to sets of denial constraints.

One can exploit the connection between causes and repairs to understand the computational complexity of the former by leveraging existing results for the latter. Beyond the fact that computing or deciding actual causes can be done in polynomial time in data for CQs and UCQs [5,36], one can show that most computational problems related to responsibility are hard, because they are also hard for repairs, in particular, for C-repairs (all this in data complexity) [32]. In particular, one can prove [5]: (a) The *responsibility problem*, about deciding if a tuple has responsibility above a certain threshold, is *NP*-complete for UCQs. (b) Computing $\rho_D(\tau)$ is $FP^{NP(log(n))}$-complete for BCQs. This the *functional*, non-decision, version of the responsibility problem. The complexity class involved is that of computational problems that use polynomial time with a logarithmic number of calls to an oracle in *NP*. (c) Deciding if a tuple τ is a most responsible cause is $P^{NP(log(n))}$-complete for BCQs. The complexity class is as the previous one, but for decision problems [1].

4.2 Repair-Programs for Causality in Databases

Answer-set programs (ASPs) can be used to specify, compute and query S- and C-repairs. These ASPs are called "repair programs". We will show the main ideas behind them by means of an example. For a more complete treatment see [4,17].

Example 8 (Example 2 cont.). Let us consider the DC associated to the query Q in (2): $\kappa(Q) : \neg \exists x \exists y (S(x) \wedge R(x,y) \wedge S(y))$.

The given database is inconsistent w.r.t. $\kappa(Q)$, and we may consider its repairs. Its *repair program* contains the D as set of facts, now (only for convenience) with global tuple identifiers (tids) in the first attribute: $R(1,a,b), R(2,c,d), R(3,b,b), S(4,a), S(5,c), S(6,b)$.

The main rule is the properly *repair rule*:

$$S'(t_1,x,\mathsf{d}) \vee R'(t_2,x,y,\mathsf{d}) \vee S'(t_3,y,\mathsf{d}) \longleftarrow S(t_1,x), R(t_2,x,y), S(t_3,y).$$

Here, d is an annotation constant for "tuple deleted". This rule detects in its body (its right-hand side) a violation of the DC. If this happens, its head (its left-hand-side) instructs the deletion of one of the tuples participating in the violation. The semantics of the program forces the choice of only one atom in the head (unless forced otherwise by other rules in the program, which does not occur in repair programs). Different choices will lead to different models of the program, and then, to different repairs.

In order to "build" the repairs, we need the *collection rules*:

$$S'(t,x,\mathsf{s}) \longleftarrow S(t,x), \; not \; S'(t,x,\mathsf{d}). \quad \text{etc.}$$

Here, s is an annotation for "tuple stays in repair"; and the rule collects the tuples in the original instance that have not been deleted.

There is a one-to-one correspondence between the answer-sets of the repair program and the database repairs. Actually, a model M of the program determines an S-repair D' of D, as $D' := \{R(\bar{c}) \mid R'(t, \bar{c}, \mathsf{s}) \in M\}$. Conversely, every S-repair can obtained in this way.

In this example, the S-repair, $D_1 = \{R(a, b), R(c, d), R(b, b), S(a), S(c)\}$, can be obtained from the model $M_1 = \{R'(1, a, b, \mathsf{s}), R'(2, c, d, \mathsf{s}), R'(3, b, b, \mathsf{s}), S'(4, a, \mathsf{s}), S'(5, c, \mathsf{s}), S'(6, b, \mathsf{d}), \ldots\}$. Actually, D_1 is a C-repair.

There is another S-repair, $D_2 = \{R(c, d), S(a), S(c), S(b)\}$, that is associated to the model $M_2 = \{R'(1, a, b, \mathsf{d}), R'(2, c, d, \mathsf{s}), R'(3, b, b, \mathsf{d}), S'(4, a, \mathsf{s}), S'(5, c, \mathsf{s}), S'(6, b, \mathsf{s}), \ldots\}$. This is not a C-repair. $\qquad\square$

For sets of DCs, repair programs can be made *normal*, i.e. non-disjunctive [17]. CQA becomes query answering under the *cautious or skeptical semantics* of ASPs (i.e. true in *all* repairs), which, for normal programs, is *NP*-complete (in data). This matches the data complexity of consistent QA under DCs (c.f. [4] for references to complexity of CQA).

Now, if we want to obtain from the program only those models that correspond to C-repairs, we can add *weak program constraints* (WCs), as shown in the example.

Example 9 (Example 8 cont.). Let us add to the program the WCs

$$:\sim\; R(t, \bar{x}), R'(t, \bar{x}, \mathsf{d})$$
$$:\sim\; S(t, \bar{x}), S'(t, \bar{x}, \mathsf{d}).$$

A *(hard) program constraint* in a program [29], usually denoted as

$$:-\; P_1(\bar{x}_1), \ldots, P_1(\bar{x}_n),$$

leads to discarding all the models where the join in the RHS of the constraint holds. Weak program constraints, now preceded by a ":$\sim$", may be violated by a model, buy only the models where the *number* of violations of them is minimized are kept. In our example, the WCs have the effect of minimizing the number of deleted tuples. In this way, we obtain as models only C-repairs.

In our example, we obtain C-repair D_1, corresponding to model M_1, but not S-repair D_2, because it is associated to model M_2 that is discarded due to the WCs. $\qquad\square$

As we already mentioned, C-repairs are those that can be used to obtain most-responsible actual causes. Accordingly, the latter task can be accomplished through the use of repair programs with weak constraints. We illustrate this by means of our example (c.f. [8] for a detailed treatment). Actually, cause and responsibility computation become query answering on extended repair programs. In them, causes will be represented by means of the tids we introduced for repair programs.

Example 10 (Example 9 cont.). The causes can be obtained through a new predicate, defined by the rules

$$Cause(t) \longleftarrow R'(t, x, y, \mathsf{d}),$$
$$Cause(t) \longleftarrow S'(t, x, \mathsf{d}),$$

because they correspond to deleted tuples in a repair. If we want to obtain them, it is good enough to pose a query under the *brave semantics*, which returns what is true in *some* model: $\Pi \models_{brave} Cause(t)$?

However, we would like to obtain contingency sets (for causes) and responsibilities. We will concentrate on maximum-responsibility causes and their (maximum) responsibilities, for which we assume the repair program has weak constraints, as above (c.f. [8] for non-maximum responsibility causes).

We first introduce a new binary predicate, to collect a cause and an associated contingency tuple (which is deleted together with the tuple-cause in a same repair). This predicate is of the form $CauCon(t, t')$, indicating that t is actual cause, and t' is a member of the former's contingency set. For this, for each pair of predicates P_i, P_j, not necessarily different, in the DC $\kappa(\mathcal{Q})$, we introduce the rule:

$$CauCon(t, t') \longleftarrow P'_i(t, \bar{x}_i, \mathsf{d}), \ P'_j(t', \bar{x}_j, \mathsf{d}), \ t \neq t'.$$

This will make t' a member of t's contingency set. In our example, we have the rule:

$$CauCon(t, t') \longleftarrow S'(t, x, \mathsf{d}), R'(t', u, v, \mathsf{d}),$$

where the inequality is not needed (for having different predicates), but also, among others,

$$CauCon(t, t') \longleftarrow S'(t, x, \mathsf{d}), S'(t', u, \mathsf{d}), t \neq t'.$$

In model M_1, corresponding to C-repair D_1, where there is no pair of simultaneously deleted tuples, we have no $CauCon$ atoms. Had model M_2 not been discarded due to the WCs, we would find in it (actually in its extension) the atoms: $CauCon(1, 3)$ and $CauCon(3, 1)$. □

Contingency sets, which is what we want next, are *sets*, which in general are not returned as objects from an ASP. However, there are extensions of ASP and their implementations, such as *DLV* [29], that, trough aggregations, support set construction. This is the case of *DLV-Complex* [15,16], that we have used in for running repair programs and their extensions. We do this as follows (in the program below, t, t' are variables).

$$preCon(t, \{\}) \leftarrow Cause(t), \ not \ Aux_1(t) \tag{6}$$
$$Aux_1(t) \leftarrow CauCon(t, t') \tag{7}$$
$$preCon(t, \{t'\}) \leftarrow CauCon(t, t') \tag{8}$$
$$preCon(t, \#union(C, \{t''\})) \leftarrow CauCon(t, t''), preCon(t, C), \tag{9}$$
$$not \ \#member(t'', C)$$
$$Con(t, C) \leftarrow preCon(t, C), not \ Aux_2(t, C) \tag{10}$$
$$Aux_2(t, C) \leftarrow CauCon(t, t'), \#member(t', C)$$

The auxiliary predicate in rule (2) is used to avoid a non-safe negation. That predicate is defined by rule (7). We are capturing here causes that do not have contingency companions, and then, they have an empty contingency set. Rule (8) is indeed redundant, but shows the main idea: a contingency companion of a cause is taken as element into the latter's pre-contingency set. In rule (10) we have an auxiliary predicate for the same reason as in the first rule. The main idea is to stepwise keep adding by means of set union (c.f. rule (9)), a contingent element to a possibly partial contingency set, until there is nothing left to add. These maximal contingency sets are obtained with rule (10).

In each model of the program with WCs, these contingency sets will have the same minimum size, and will lead to maximum responsibility causes. Responsibility computation can be done, with numerical aggregation supported by *DLV-Complex*, as follows:

$$pre\text{-}rho(t, n) \leftarrow \#count\{t' : CauCon(t, t')\} = n$$
$$rho(t, m) \leftarrow m * (pre\text{-}rho(t, m) + 1) = 1$$

The first rule gives us the (minimum) size, n, of contingency sets, which leads to a responsibility of $\frac{1}{1+n}$. The responsibility of a (maximum responsibility) cause t can be obtained through a query to the extended program: $\Pi^e \models_{brave} rho(t, X)?$.

ASP with WCs computation has exactly the required expressive power or computational complexity needed for maximum-responsibility computation [8].

4.3   The Example with *DLV-Complex*

In this section we show in detail the running example in Sect. 4.2, fully specified and executed with the *DLV-Complex* system [15,16]. C.f. [8] for more details.

Example 11 (Example 8 cont.). The first fragment of the *DLV* program below, shows facts for database D, and the disjunctive repair rule for the DC $\kappa(Q)$. In it, and in the rest of this section, R_a, S_a, ... stand for $R', S', ...$ used before, with the subscript _a for "auxiliary". We recall that the first attribute of a predicate holds a variable or a constant for a tid; and the last attribute of R_a, etc. holds an annotation constant, d or s, for "deleted" (from the database) or "stays" in a repair, resp. (In *DLV* programs, variables start with a capital letter, and constants, with lower-case).

```
R(1,a,b). R(2,c,d). R(3,b,b). S(4,a). S(5,c). S(6,b).

S_a(T1,X,d) v R_a(T2,X,Y,d) v S_a(T3,Y,d) :- S(T1,X),R(T2,X,Y), S(T3,Y).
S_a(T,X,s)    :- S(T,X), not S_a(T,X,d).
R_a(T,X,Y,s) :- R(T,X,Y), not R_a(T,X,Y,d).
```

DLV returns the stable models of the program, as follows:

```
{S_a(6,b,d), R_a(1,a,b,s), R_a(2,c,d,s), R_a(3,b,b,s),
 S_a(4,a,s), S_a(5,c,s)}
```

```
{R_a(1,a,b,d), R_a(3,b,b,d), R_a(2,c,d,s), S_a(4,a,s),
 S_a(5,c,s), S_a(6,b,s)}

{S_a(4,a,d), R_a(3,b,b,d), R_a(1,a,b,s), R_a(2,c,d,s),
 S_a(5,c,s), S_a(6,b,s)}
```

These three stable models (that do not show here the original EDB) are associated to the S-repairs D_1, D_2, D_3, resp. Only tuples with tids $1, 3, 4, 6$ are at some point deleted. In particular, the first model corresponds to the C-repair

$$D_1 = \{R(s_4, s_3), R(s_2, s_1), R(s_3, s_3), S(s_4), S(s_2)\}.$$

Now, to compute causes and their accompanying deleted tuples we add to the program the rules defining *Cause* and *CauCont*:

```
        cause(T) :- S_a(T,X,d).
        cause(T) :- R_a(T,X,Y,d).
 cauCont(T,TC) :- S_a(T,X,d), S_a(TC,U,d), T != TC.
 cauCont(T,TC) :- R_a(T,X,Y,d), R_a(TC,U,V,d), T != TC.
 cauCont(T,TC) :- S_a(T,X,d), R_a(TC,U,V,d).
 cauCont(T,TC) :- R_a(T,X,Y,d), S_a(TC,U,d).
```

Next, contingency sets can be computed by means of *DLV-Complex*, on the basis of the rules defining predicates *cause* and *cauCont* above:

```
        preCont(T,{TC}) :- cauCont(T,TC).
 preCont(T,#union(C,{TC})) :- cauCont(T,TC), preCont(T,C),
                              not #member(TC,C).
        cont(T,C)    :- preCont(T,C), not HoleIn(T,C).
        HoleIn(T,C)  :- preCont(T,C), cauCont(T,TC),
                              not #member(TC,C).
        tmpCont(T)   :- cont(T,C), not #card(C,0).
        cont(T,{})   :- cause(T), not tmpCont(T).
```

The last two rules associate the empty contingency set to counterfactual causes.

The three stable models obtained above will now be extended with *cause*- and *cont*-atoms, among others (unless otherwise stated, *preCont*-, *tmpCont*-, and *HoleIn*-atoms will be filtered out from the output); as follows:

```
{S_a(4,a,d), R_a(3,b,b,d), R_a(1,a,b,s), R_a(2,c,d,s),
 S_a(5,c,s), S_a(6,b,s), cause(4), cause(3), cauCont(4,3),
 cauCont(3,4), cont(3,{4}), cont(4,{3})}

{R_a(1,a,b,d), R_a(3,b,b,d), R_a(2,c,d,s), S_a(4,a,s),
 S_a(5,c,s), S_a(6,b,s), cause(1), cause(3), cauCont(1,3),
 cauCont(3,1), cont(1,{3}), cont(3,{1})}

{S_a(6,b,d), R_a(1,a,b,s), R_a(2,c,d,s), R_a(3,b,b,s),
 S_a(4,a,s), S_a(5,c,s), cause(6), cont(6,{})}
```

The first two models above show tuple 3 as an actual cause, with one contingency set per each of the models where it appears as a cause. The last line of the third model shows that cause (with tid) 6 is the only counterfactual cause (its contingency set is empty).

The responsibility ρ can be computed via predicate $preRho(T, N)$ that returns $N = \frac{1}{\rho}$, that is the inverse of the responsibility, for each tuple with tid T *and local to a model* that shows T as a cause. We concentrate on the computation of $preRho$ in order to compute with integer numbers, as supported by *DLV-Complex*, which requires setting an upper integer bound by means of maxint, in this case, at least as large as the largest tid:

```
#maxint = 100.
preRho(T,N + 1) :- cause(T), #int(N), #count{TC: cauCont(T,TC)} = N.
```

where the local (pre)responsibility of a cause (with tid) T within a repair is obtained by counting how many instances of $cauCont(T, ?)$ exist in the model, which is the size of the local contingency set for T plus 1. We obtain the following (filtered) output:

```
{S_a(4,a,d), R_a(3,b,b,d), cause(4), cause(3),
   preRho(3,2), preRho(4,2), cont(3,{4}), cont(4,{3})}

{R_a(1,a,b,d), R_a(3,b,b,d), cause(1), cause(3),
   preRho(1,2), preRho(3,2), cont(1,{3}), cont(3,{1})}

{S_a(6,b,d), cause(6), preRho(6,1), cont(6,{})}
```

The first model shows causes 3 and 4 with a pre-rho value of 2. The second one, causes 3 and 1 with a pre-rho value of 2. The last model shows cause 6 with a pre-rho value of 1. This is also a maximum-responsibility cause, actually associated to a C-repair. Inspecting the three models, we can see that the overall pre-responsibility of cause 3 (the minimum of its pre-rho values) is 2, similarly for cause 1. For cause 6 the overall pre-responsibility value is 1.

Now, if we want only maximum-responsibility causes, we add weak program constraints to the program above, to minimize the number of deletions:

```
:~ S_a(T,X,d).
:~ R_a(T,X,Y,d).
```

DLV shows only repairs with the least number of deletions, in this case:

```
Best model: {S_a(6,b,d), R_a(1,a,b,s), R_a(2,c,d,s), R_a(3,b,b,s),
             S_a(4,a,s), S_a(5,c,s), cause(6), preRho(6,1), cont(6,{})}
Cost ([Weight:Level]): <[1:1]>
```

As expected, only repair D_1 is obtained, where only $S(6, s_3)$ is a cause, and with responsibility 1, making it a maximum-responsibility cause.   □

5   Causal Explanations in Databases: Attribute-Level

In Sect. 4.1 we saw that: (a) there are different database repair-semantics; and (b) tuples as causes for query answering can be obtained from S- and C-repairs. We can extrapolate from this, and *define*, as opposed to only reobtain, notions of causality on the basis of a repair semantics. This is what we will do next in order to define attribute-level causes for query answering in databases.

We may start with a repair-semantics S for databases under, say denial constraints (this is the case we need here, but we could have more general ICs). Now, we have a database D and a true BCQ Q. As before, we have an associated (and violated) denial constraint $\kappa(Q)$. There will be S-repairs, i.e. sanctioned as such by the repair semantics S. More precisely, the repair-semantics S identifies a class $Rep^S(D, \kappa(Q))$ of admissible and consistent instances that "minimally" depart from D. On this basis, S-causes can be defined as in Sect. 4.1(a)–(b). Of course, "minimality" has to be defined, and comes with S.

We will develop this idea, at the light of an example, with a particular repair-semantics, and we will apply it to define attribute-level causes for query answering, i.e. we are interested in attribute values in tuples rather than in whole tuples. The repair semantics we use here is natural, but others could be used instead.

Example 12. Consider the database D, with tids, and query Q: $\exists x \exists y (S(x) \wedge R(x, y) \wedge S(y))$, of Example 2 and the associated denial constraint $\kappa(Q)$: $\neg \exists x \exists y (S(x) \wedge R(x, y) \wedge S(y))$.

R	A	B
t_1	a	b
t_2	c	d
t_3	b	b

S	C
t_4	a
t_5	c
t_6	b

Since $D \not\models \kappa(Q)$, we need to consider repairs of D w.r.t. $\kappa(Q)$.

Repairs will be obtained by "minimally" changing attribute values by NULL, as in SQL databases, which cannot be used to satisfy a join. In this case, minimality means that *the set* of values changed by NULL is minimal under set inclusion. These are two different minimal-repairs:

R	A	B
t_1	a	b
t_2	c	d
t_3	b	b

S	C
t_4	a
t_5	c
t_6	NULL

R	A	B
t_1	a	NULL
t_2	c	d
t_3	b	NULL

S	C
t_4	a
t_5	c
t_6	b

It is easy to check that they do not satisfy $\kappa(Q)$. If we denote the changed values by the tid with the position where the change occurred, then the first repair is characterized by the set $\{t_6[1]\}$, whereas the second, by the set $\{t_1[2], t_3[2]\}$. Both are minimal since none of them is contained in the other.

Now, we could also introduce a notion of *cardinality-repair*, keeping those where the number of changes is a minimum. In this case, the first repair qualifies, but not the second.

These repairs identify (actually, define) the value in $t_6[1]$ as a maximum-responsibility cause for $\mathcal{Q}$ to be true (with responsibility 1). Similarly, $t_1[2]$ and $t_3[2]$ become actual causes, that do need contingent companion values, which makes them take a responsibility of $\frac{1}{2}$ each.   □

We should emphasize that, under this semantics, we are considering attribute values participating in joins as interesting causes. A detailed treatment can be found in [8]. Of course, one could also consider as causes other attribute values in a tuple that participate in a query (being true), e.g. that in $t_3[1]$, but making them *non-prioritized* causes. One could also think of adjusting the responsibility measure in order to give to these causes a lower score.

5.1 ASPs for Attribute-Level Causality

So as in Sects. 4.2 and 4.3, we can specify attribute-level causes via attribute-based repairs, and their ASPs. We show this at the light of an example that is given directly using *DLV* code (c.f. [8] for more details).

Example 13. Consider the database instance

$$D = \{S(a), S(b), R(b, c), R(b, d), R(b, e)\},$$

and the BCQ $\mathcal{Q} \colon \exists x \exists y(S(x) \wedge R(x, z))$, which is true in D, and for which we want to find attribute-level causes.

We consider the DC corresponding to the negation of query, namely

$$\kappa \colon \ \neg \exists x \exists y(S(x) \wedge R(x, z)).$$

Since $D \not\models \kappa$, D is inconsistent. The updated instance

$$D_2 = \{S(a), S(\mathsf{NULL}), R(b, c), R(b, d), R(b, e)\}$$

is consistent (among others obtained by updates with NULL), i.e. $D_2 \models \kappa$.

In the *DLV* program below, R_a, and S_a are the auxiliary predicates associated to R and S. They accommodate annotation constants in their last argument. The annotation constants tr, u, fu and s stand for "in transition" (i.e. initial or updated tuple, that could be further updated), "has been updated", "is final update", and "stays in repair", resp. The tuples already contain tuple-ids. Here, T, T2, X, Y, ... are variables.

```
S(1,a).  S(2,b).  R(3,b,c).  R(4,b,d).  R(5,b,e).

  S_a(T,X,tr)    :- S(T,X).
  S_a(T,X,tr)    :- S_a(T,X,u).
  R_a(T,X,Y,tr)  :- R(T,X,Y).
  R_a(T,X,Y,tr)  :- R_a(T,X,Y,u).
```

This part of the program so far provides, as facts, the tuples in the database with their tids. It also defines each of these tuples as "in transition". The same for those that have been updated.

The updates themselves come in the following portion of the program. In it, null is treated as any other constant, and can be compared with other constants (as opposed to their occurrence as NULL in SQL, where any comparison involving it is considered to be false).

The first two rules capture, in the first three atoms in the body, a violation of the constraints, i.e. a join through a non-null value, for X. The last atom in the body of the first rule says that the value for X in R is not updated to *null*, then, as specified in the head of the rule, it has to be updated in S. The second rule is similar, but the other way around.[2]

```
S_a(T,null,u) :- S_a(T,X,tr), R_a(T2,X,Y,tr), X != null,
                 not R_a(T2,null,Y,u).
R_a(T,null,Y,u) :- R_a(T,X,Y,tr), S_a(T2,X,tr), X != null,
                 not S_a(T2,null,u).
```

In R_a(t,m,n,fu) below, annotation fu means that the atom with tid t has reached its final update (during the program evaluation). In particular, R(t,m,n) has already been updated, and annotation u should appear in the new, updated atom, say R_a(t,m1,n1,u), and this tuple cannot be updated any further (because relevant updateable attribute values have already been replaced by null if necessary). This is captured by the next five rules:

```
S_a(T,X,fu) :- S_a(T,X,u), not auxS1(T,X).
auxS1(T,X) :- S(T,X), S_a(T,null,u), X != null.

R_a(T,X,Y,fu) :- R_a(T,X,Y,u), not auxR1(T,X,Y), not auxR2(T,X,Y).
auxR1(T,X,Y) :- R(T,X,Y), R_a(T,null,Y,u), X != null.
auxR2(T,X,Y) :- R(T,X,Y), R_a(T,X,null,u), Y != null.
```

The final six rules collect what stays in a repair, as annotated with s:

```
S_a(T,X,s) :- S_a(T,X,fu).
S_a(T,X,s) :- S(T,X), not auxS(T).
   auxS(T) :- S_a(T,X,u).
R_a(T,X,Y,s) :- R_a(T,X,Y,fu).
R_a(T,X,Y,s) :- R(T,X,Y), not auxR(T).
   auxR(T) :- R_a(T,X,Y,u).
```

Two stable models are returned, corresponding to two attribute-based repairs: (we skip the atoms without annotation s)

---

[2] Those two *normal* rules could be replaced by a single disjunctive rule:
$S\_a(T, null, u) \lor R\_a(T, null, Y, u) \leftarrow S\_a(T, X, tr), R\_a(T2, X, Y, tr), X \neq null$. For this kind of disjunctive repair programs one can show that the normal and disjunctive versions are equivalent, i.e. they have the same models. This is because, the disjunctive program becomes *head-cycle free* [20].

```
{S_a(1,a,s), S_a(2,b,s), R_a(3,null,c,s), R_a(5,null,e,s), R_a(4,null,d,s)}
{S_a(1,a,s), R_a(3,b,c,s), R_a(4,b,d,s), R_a(5,b,e,s), S_a(2,null,s)}
```

The second model corresponds to the repair D_2 given at the beginning of this example.

We could extend the program with rules to collect the attribute values that are causes for the query to be true:

```
cause(T,1,X) :- R(T,X,Y), R_a(T,null,Z,s).
cause(T,2,Y) :- R(T,X,Y), R_a(T,Z,null,s).
cause(T,1,X) :- S(T,X), S_a(T,null,s).
```

Here, the second argument indicates the position where the cause, as a value, appears in a tuple. Remember that the tids are global, so having them in the first body atom in these rules will always make these rules to be evaluated with different tids, which come from the original database.

Here, we are assuming the original database does not have nulls. If it does, it is good enough to add the extra condition X != null in the body of the first rule, and similarly for the other rules. Each model will return some causes. If we want them all, and we have no interest in the repairs or the complete models, we can just pose a query under the *brave semantics*:   :- cause(U,V,W)? We will obtain all the cause-atoms that appear in *some* of the models of the extended program, e.g. cause(3,1,b), i.e. the value b in the first attribute, "1", of tuple with id 3.   □

6   Causes Under Integrity Constraints

In this section we consider tuples as causes for query answering in the more general setting where databases are subject to integrity constraints (ICs). In this scenario, and in comparison with Sect. 3.1, not every intervention on the database is admissible, because the ICs have to be satisfied. As a consequence, the definitions of cause and responsibility have to be modified accordingly. We illustrate the issues by means of an example. More details can be found in [6,8].

We start assuming that a database D satisfies a set of ICs, Σ, i.e. $D \models \Sigma$. If we concentrate on BCQs, or more, generally on monotone queries, and consider causes at the tuple level, only instances obtained from D by interventions that are tuple deletions have to be considered; and they should satisfy the ICs. More precisely, for τ to be actual cause for Q, with a contingency set Γ, it must hold [6]:

(a) $D \smallsetminus \Gamma \models \Sigma$, and $D \smallsetminus \Gamma \models Q$.
(b) $D \smallsetminus (\Gamma \cup \{\tau\}) \models \Sigma$, and $D \smallsetminus (\Gamma \cup \{\tau\}) \not\models Q$.

The *responsibility* of τ, denoted $\rho_{Q(\bar{a})}^{D,\Sigma}(\tau)$, is defined as in Sect. 3.1, through minimum-size contingency sets.

Example 14. Consider the database instance D as below, initially without additional ICs.

Dep	DName	TStaff
t_1	Computing	John
t_2	Philosophy	Patrick
t_3	Math	Kevin

Course	CName	TStaff	DName
t_4	COM08	John	Computing
t_5	Math01	Kevin	Math
t_6	HIST02	Patrick	Philosophy
t_7	Math08	Eli	Math
t_8	COM01	John	Computing

Let us first consider the following open query: (The fact that it is open is not particularly relevant, because we can instantiate the query with the answer, obtaining a Boolean query).

$$\mathcal{Q}(x)\colon \exists y \exists z (Dep(y, x) \wedge Course(z, x, y)). \tag{11}$$

In this case, we get answers other that *yes* or *no*. Actually, $\langle \text{John} \rangle \in \mathcal{Q}(D)$, the set of answers to $\mathcal{Q}$, and we look for causes for this particular answer. It holds: (a) t_1 is a counterfactual cause; (b) t_4 is actual cause with single minimal contingency set $\Gamma_1 = \{t_8\}$; (c) t_8 is actual cause with single minimal contingency set $\Gamma_2 = \{t_4\}$.

Let us now impose on D the *inclusion dependency* (IND):

$$\psi\colon \quad \forall x \forall y \, (Dep(x, y) \rightarrow \exists u \, Course(u, y, x)), \tag{12}$$

which is satisfied by D. Now, t_4 t_8 are not actual causes anymore; and t_1 is still a counterfactual cause.

Let us now consider the query

$$\mathcal{Q}_1(x)\colon \exists y \, Dep(y, x). \tag{13}$$

Now, $\langle \text{John} \rangle \in \mathcal{Q}_1(D)$, and under the IND (12), we obtain the same causes as for $\mathcal{Q}$, which is not surprising considering that $\mathcal{Q} \equiv_\psi \mathcal{Q}_1$, i.e. the two queries are logically equivalent under (12).

And now, consider the query:

$$\mathcal{Q}_2(x)\colon \exists y \exists z \, Course(z, x, y), \tag{14}$$

for which $\langle \text{John} \rangle \in \mathcal{Q}_2(D)$.

For this query we consider the two scenarios, with and without imposing the IND. Without imposing (12), t_4 and t_8 are the only actual causes, with contingency sets $\Gamma_1 = \{t_8\}$ and $\Gamma_2 = \{t_4\}$, resp.

However, imposing (12), t_4 and t_8 are still actual causes, but we lose their smallest contingency sets Γ_1 and Γ_2 we had before: $D \smallsetminus (\Gamma_1 \cup \{t_4\}) \not\models \psi$, $D \smallsetminus (\Gamma_2 \cup \{t_8\}) \not\models \psi$. Actually, the smallest contingency set for t_4 is $\Gamma_3 = \{t_8, t_1\}$; and for t_8, $\Gamma_4 = \{t_4, t_1\}$.

We can see that under the IND, the responsibilities of t_4 and t_8 decrease: $\rho^D_{\mathcal{Q}_2(\text{John})}(t_4) = \frac{1}{2}$, but $\rho^{D,\psi}_{\mathcal{Q}_2(\text{John})}(t_4) = \frac{1}{3}$. Tuple t_1 is not an actual cause, but it affects the responsibility of actual causes. □

Some results about causality under ICs can be obtained [6]: (a) Causes are preserved under logical equivalence of queries under ICs, (b) Without ICs, deciding causality for BCQs is tractable, but their presence may make complexity grow. More precisely, there are a BCQ and an inclusion dependency for which deciding if a tuple is an actual cause is *NP*-complete in data.

6.1   Specifying and Computing Causes Under Integrity Constraints

ASPs for computation of causes and responsibilities under ICs can be produced. However, Example 14 shows that contingency sets may be affected by the presence of ICs.

Example 15 (Example 14 cont.).   Database D violates the DC κ_2 : $\neg \exists z\, Course(z, \mathsf{John})$ associated to query $\mathcal{Q}_2$ and its answer John. Without considering ψ, its only minimal repair is $D' = D \smallsetminus \{\tau_4, \tau_8\}$. However, if we accept minimal repairs that also satisfy ψ (when D already did), then the only minimal repair is $D'' = D \smallsetminus \{\tau_1, \tau_4, \tau_8\}$.   □

This example shows that, in the presence of a set of hard ICs Ψ, the repairs w.r.t. to another set of ICs Σ that also satisfy Ψ may not be among the repairs w.r.t. Σ without consideration for Ψ. So, it is not only a matter of discarding some of the unwanted repairs w.r.t. Σ alone.

The example also shows that, in the presence of a hard set of ICs Ψ, the characterization of causes in terms of repairs (as in Sect. 3.1) has to be revised. Doing this should be relatively straightforward for repairs of D w.r.t. the DCs Σ that have origin in UBCQs, and are maximally contained in D under set-inclusion, and also satisfy the hard constraints Ψ. Instead of giving a general approach, we show how a repair-program could be used to reobtain the results obtained in Example 14, where an inclusion dependency is our IC.

Example 16 (Examples 14 and 15 cont.). Without considering the IC ψ, the repair-program for D w.r.t. the DC κ_2 is:

1. The extensional database as a set of facts corresponding to the table. For example, $Dept(1, \mathsf{computing}, \mathsf{john})$, etc.
2. Repair rule for κ_2: $Course'(t, z, \mathsf{john}, \mathsf{d}) \leftarrow Course(t, z, \mathsf{john})$.
3. Persistence rule: $Course'(t, x, y, \mathsf{s}) \leftarrow Course(t, x, y),\ not\ Course'(t, x, y, \mathsf{d})$.

We have to add to this program, rules that take care of repairing w.r.t. ψ in case it is violated via deletions from $Course$:

4. $Dept'(t', x, y, \mathsf{d}) \leftarrow Dept(t', x, y), not\ aux(y)$
5. $aux(y) \leftarrow Course'(t, x, y, \mathsf{s})$.
6. $Dept'(t, x, y, \mathsf{s}) \leftarrow Dept(t, x, y),\ not\ Dept'(t, x, y, \mathsf{d})$.

Notice that violations of the inclusion dependency that may arise from deletions from $Course$ are being repaired through deletions from $Dept$. The only stable model of this program corresponds to the repair in Example 15.   □

Notice that the definition of actual cause under ICs opens the ground for a definition of a notion of *underlying* (*hidden, latent*) cause. In Example 14, τ_1 could be such a cause. It is not strictly an actual cause, but it has to appear in every minimal contingency set. Similarly, Example 15 shows that τ_1 has to appear in the difference between the original instance and every minimal repair.

7   Measuring Database Inconsistency and ASPs

A database D is expected to satisfy a given set of integrity constraints (ICs), Σ, that come with the database schema. However, databases may be inconsistent in that those ICs are not satisfied. A natural question is: *To what extent, or how much inconsistent is D w.r.t. Σ, in quantitative terms?*. This problem is about defining a *global numerical score* for the database, to capture its "degree of inconsistency". This number can be interesting *per se*, as a measure of data quality (or a certain aspect of it), and could also be used to compare two databases (for the same schema) w.r.t. (in)consistency.

Scores for individual tuples in relation to their contribution to inconsistency can be obtain through responsibility scores for query answering, because every IC gives rise to a violation view; and a tuple contained in it can be scored. Also Shapley values can be applied (c.f. Sect. 8; see also [31]).

Inconsistency measures have been introduced and investigated in knowledge representation, but mainly for propositional theories; and, in the first-order case through grounding. In databases, it is more natural to consider the different nature of the combination of a database, as a structure, and ICs, as a set of first-order formulas. It is also important to consider the asymmetry: databases are inconsistent or not, not the combination. Furthermore, the relevant issues that are usually related to data management have to do with algorithms and computational complexity; actually, in terms of the database and its size. Notice that ICs are usually few and fixed, whereas databases can be huge.

In [7], a particular and natural *inconsistency measure* (IM) was introduced and investigated. Maybe more important than the particular measure, the research program to be developed around such an IM is particularly relevant. More specifically, the measure was inspired by one used for functional dependencies (FDs), and reformulated and generalized *in terms of a class of database repairs*. In addition to algorithms, complexity results, approximations for hard cases of IM computation, and the dynamics of the IM under updates, ASPs were proposed for the computation of this measure. We concentrate on this part in the rest of this section. We use the notions and notation introduced in Sect. 4.1 and its Example 7.

For a database D and a set of *denial constraints* Σ (this is not essential, but to fix ideas), we have the classes of subset-repairs (or S-repairs), and cardinality-repairs (or C-repairs), denoted $Srep(D, \Sigma)$ and $Crep(D, \Sigma)$, resp. The following IMs are introduced:

$$inc\text{-}deg^S(D, \Sigma) := \frac{|D| - max\{|D'| \ : \ D' \in Srep(D, \Sigma)\}}{|D|},$$

$$inc\text{-}deg^C(D, \Sigma) := \frac{|D| - max\{|D'| \ : \ D' \in Crep(D, \Sigma)\}}{|D|}.$$

We can see that it is good enough to concentrate on $inc - deg^C(D, \Sigma)$ since it gives the same value as $inc - deg^S(D, \Sigma)$. Actually, to compute it, one C-repair is good enough. It is clear that $0 \leq inc\text{-}deg^C(D, \Sigma) \leq 1$, with value 0 when D

consistent. Notice that one could use other repair semantics instead of C-repairs [7].

Example 17 (Example 7 cont.). Here, $Srep(D, \Sigma) = \{D_1, D_2\}$ and $Crep(D, \Sigma) = \{D_1\}$. It holds: $inc - deg^S(D, \Sigma) = \frac{4-|D_1|}{4} = inc - deg^C(D, \Sigma) = \frac{4-|D_1|}{4} = \frac{1}{4}$. □

The complexity of computing $inc - deg^C(D, \Sigma)$ for DCs belongs to $FP^{NP(log(n))}$, in data complexity. Furthermore, there is a relational schema and a set of DCs Σ for which computing $inc - deg^C(D, \Sigma)$ is $FP^{NP(log(n))}$-complete.

It turns out that complexity and efficient computation results can be obtained via C-repairs, and we end up confronting graph-theoretic problems. Actually, C-repairs are in one-to-one correspondence with maximum-size independent sets in hypergraphs [32].

Example 18. Consider the database $D = \{A(a), B(a), C(a), D(a), E(a)\}$, which is inconsistent w.r.t. the set of DS:

$$\Sigma = \{\neg \exists x (B(x) \wedge E(x)), \ \neg \exists x (B(x) \wedge C(x) \wedge D(x)), \ \neg \exists x (A(x) \wedge C(x))\}.$$

We obtain the following *conflict hyper-graph* (CHG), where tuples are the nodes, and a hyperedge connects tuples that together violate a DC:

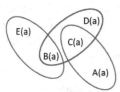

S-repairs are maximal independent sets: $D_1 = \{B(a), C(a)\}$, $D_2 = \{C(a), D(a), E(a)\}$, $D_3 = \{A(a), B(a), D(a)\}$; and the C-repairs are D_2, D_3. □

There is a connection between C-repairs and hitting-sets (HS) of the hyperedges of the CHG: The removal from D of the vertices in a minimum-size HS produces a C-repair. The connections between hitting-sets in hypergraphs and C-repairs can be exploited for algorithmic purposes, and to obtain complexity and approximation results [7].

It turns out that the IM can be computed via ASPs, and not surprisingly by now, via specification of C-repairs.

Example 19 (Example 12 cont.). Consider the following DC and database (with tids)

$$\kappa: \quad \neg \exists x \exists y (S(x) \wedge R(x, y) \wedge S(y)),$$
$$D = \{R(1, a, b), R(2, c, d), R(3, b, b), S(4, a), S(5, c), S(6, b)\}.$$

The repair-ASP specifying C-repairs contains the DB D, plus the rules:

$$S'(t_1, x, \mathsf{d}) \vee R'(t_2, x, y, \mathsf{d}) \vee S'(t_3, y, \mathsf{d}) \leftarrow S(t_1, x), R(t_2, x, y), S(t_3, y),$$
$$S'(t, x, \mathsf{s}) \leftarrow S(t, x), \ not \ S'(t, x, \mathsf{d}),$$
$$R'(t, x, y, \mathsf{s}) \leftarrow R(t, x, y), \ not \ R'(t, x, y, \mathsf{d}),$$

and weak program constraints (c.f. Example 9):

$$:\sim \quad R(\bar{x}), R'(\bar{x}, \mathsf{d}),$$
$$:\sim \quad S(\bar{x}), S'(\bar{x}, \mathsf{d}).$$

With them, we keep the models that minimize the number of deleted tuples. The C-repair D_1 is represented by the model

$$M_1 = \{R'(1, a, b, \mathsf{s}), R'(2, c, d, \mathsf{s}), R'(3, b, b, \mathsf{s}), S'(4, a, \mathsf{s}), S'(5, c, \mathsf{s}), S'(6, b, \mathsf{d}), \ldots\}.$$

Now, the IM can be computed via $|D \smallsetminus D'|$ for some (or any) C-repair D'. In this case, D_1.

With a system like *DLV-Complex*, we can specify this set difference and compute its cardinality as a simple aggregation. More precisely, we add to the program above the rules:

$$Del(t) \leftarrow S'(t, x, \mathsf{d}),$$
$$Del(t) \leftarrow R'(t, x, y, \mathsf{d}),$$
$$NumDel(n) \leftarrow \#count\{t : Del(t)\} = n.$$

The first two rules collect the tids of deleted tuples. The value for *NumDel* defined by the third rule is the number of deleted tuples (that already takes a minimum due to the weak constraints). This number is all we need to compute the IM. All the models, corresponding to C-repairs, will return the same number. For this reason, there is no need to explicitly compute all stable models, their sizes, and compare them. Actually, this value can be obtained by means of a query posed to the program: "$:- \ NumDel(x)$?", that can be answered under the brave semantics (returning answers that hold in some of the stable models). In [7, Appendix A] one can find an extended example that uses *DLV-Complex* [15, 16] for this computation.   □

8   The Shapley Value in Databases

The Shapley value was proposed in game theory by Lloyd Shapley in 1953 [45], to quantify the contribution of a player to a coalition game where players share a wealth function.[3] It has been applied in many disciplines. In particular, it has been investigated in computer science under *algorithmic game theory* [40], and it has been applied to many and different computational problems. The computation of the Shapley value is, in general, intractable. In many scenarios where it is applied its computation turns out to be $\#P$-hard [21, 22]. Here, the class $\#P$ contains the problems of *counting* the solutions for problems in *NP*. A typical problem in the class, actually, hard for the class, is $\#SAT$, about counting

---

[3] The original paper and related ones on the Shapley value can be found in the book edited by Alvin Roth [43]. Shapley and Roth shared the Nobel Prize in Economic Sciences 2012.

the number of satisfying assignments for a propositional formula. Clearly, this problem cannot be easier than SAT, because a solution for $\#SAT$ immediately gives a solution for SAT [1].

In particular, the Shapley value has been used in knowledge representation, to measure the degree of inconsistency of a propositional knowledge base [28]; in machine learning to provide explanations for the outcomes of classification models on the basis of numerical scores assigned to the participating feature values [35] (c.f. Sect. 13); and in data management to measure the contribution of a tuple to a query answer [30], which we briefly review in this section.

Consider a set of players D, and a game function, $\mathcal{G} : \mathcal{P}(D) \rightarrow \mathbb{R}$, where $\mathcal{P}(D)$ the power set of D. The Shapley value of player p in D es defined by:

$$Shapley(D, \mathcal{G}, p) := \sum_{S \subseteq D \setminus \{p\}} \frac{|S|!(|D| - |S| - 1)!}{|D|!} (\mathcal{G}(S \cup \{p\}) - \mathcal{G}(S)). \quad (15)$$

Notice that here, $|S|!(|D| - |S| - 1)!$ is the number of permutations of D with all players in S coming first, then p, and then all the others. That is, this quantity is the expected contribution of player p under all possible additions of p to a partial random sequence of players followed by a random sequence of the rests of the players. Notice the counterfactual flavor, in that there is a comparison between what happens having p vs. not having it. The Shapley value is the only function that satisfies certain natural properties in relation to games. So, it is a result of a categorical set of axioms or conditions [43].

Back to query answering in databases, the players are tuples in the database D. We also have a Boolean query $\mathcal{Q}$, which becomes a game function, as follows: For $S \subseteq D$,

$$\mathcal{Q}(S) = \begin{cases} 1 & \text{if } S \models \mathcal{Q}, \\ 0 & \text{if } S \not\models \mathcal{Q}. \end{cases}$$

With these elements we can define the Shapley value of a database tuple τ:

$$Shapley(D, \mathcal{Q}, \tau) := \sum_{S \subseteq D \setminus \{\tau\}} \frac{|S|!(|D| - |S| - 1)!}{|D|!} (\mathcal{Q}(S \cup \{\tau\}) - \mathcal{Q}(S)).$$

If the query is *monotone*, i.e. its set of answers never shrinks when new tuples are added to the database, which is the case of conjunctive queries (CQs), among others, the difference $\mathcal{Q}(S \cup \{\tau\}) - \mathcal{Q}(S)$ is always 1 or 0, and the average in the definition of the Shapley value returns a value between 0 and 1. This value quantifies the contribution of tuple τ to the query result. It was introduced and investigated in [30], for BCQs and some aggregate queries defined over CQs. We report on some of the findings in the rest of this section. The analysis has been extended to queries with negated atoms in CQs [41].

A main result obtained in [30] is about the complexity of computing this Shapley score. The following *Dichotomy Theorem* holds: For $\mathcal{Q}$ a BCQ without self-joins, if $\mathcal{Q}$ is *hierarchical*, then $Shapley(D, \mathcal{Q}, \tau)$ can be computed in polynomial-time (in the size of D); otherwise, the problem is $\#P$-complete.

Here, $\mathcal{Q}$ is hierarchical if for every two existential variables x and y, it holds: (a) $Atoms(x) \subseteq Atoms(y)$, or $Atoms(y) \subseteq Atoms(x)$, or $Atoms(x) \cap Atoms(y) = \emptyset$. For example, $\mathcal{Q} : \exists x \exists y \exists z (R(x,y) \wedge S(x,z))$, for which $Atoms(x) = \{R(x,y), S(x,z)\}$, $Atoms(y) = \{R(x,y)\}$, $Atoms(z) = \{S(x,z)\}$, is hierarchical. However, $\mathcal{Q}^{nh} : \exists x \exists y (R(x) \wedge S(x,y) \wedge T(y))$, for which $Atoms(x) = \{R(x), S(x,y)\}$, $Atoms(y) = \{S(x,y), T(y)\}$, is not hierarchical.

These are the same criteria for (in)tractability that apply to evaluation of BCQs over probabilistic databases [47]. However, the same proofs do not apply, at least not straightforwardly. The intractability result uses query $\mathcal{Q}^{nh}$ above, and a reduction from counting independent sets in a bipartite graph.

The dichotomy results can be extended to summation over CQs, with the same conditions and cases. This is because the Shapley value, as an expectation, is linear. Hardness extends to aggregates max, min, and avg over non-hierarchical queries.

For the hard cases, there is, as established in [30], an *approximation result*: For every fixed BCQ $\mathcal{Q}$ (or summation over a CQ), there is a *multiplicative fully-polynomial randomized approximation scheme* (FPRAS) [1], A, with

$$P(\tau \in D \mid \frac{Shapley(D, \mathcal{Q}, \tau)}{1 + \epsilon} \leq A(\tau, \epsilon, \delta) \leq (1 + \epsilon) Shapley(D, \mathcal{Q}, \tau)\}) \geq 1 - \delta.$$

A related and popular score, in coalition games and other areas, is the *Bahnzhaf Power Index*, which is similar to the Shapley value, but the order of players is ignored, by considering subsets of players rather than permutations thereof. It is defined by:

$$Banzhaf(D, \mathcal{Q}, \tau) := \frac{1}{2^{|D|-1}} \cdot \sum_{S \subseteq (D \setminus \{\tau\})} (\mathcal{Q}(S \cup \{\tau\}) - \mathcal{Q}(S)).$$

The Bahnzhaf-index is also difficult to compute; provably #P-hard in general. The results in [30] carry over to this index when applied to query answering. In [30] it was proved that the causal-effect score of Sect. 3.2 coincides with the Banzhaf-index, which gives to the former an additional justification.

9   Score-Based Explanations for Classification

Let us consider, as in Fig. 2, a classifier, $\mathcal{C}$, that receives as input the representation of a entity, $\mathbf{e} = \langle x_1, \ldots, x_n \rangle$, as a record of feature values, and returns as an output a label, $L(\mathbf{e})$, corresponding to the classification of input $\mathbf{e}$. In principle, we could see $\mathcal{C}$ as a black-box, in the sense that only by direct interaction with it, we have access to its input/output relation. That is, we may have no access to the mathematical classification model inside $\mathcal{C}$.

To simplify the presentation, we will assume that the classifier is *binary*, that is, for every entity $\mathbf{e}$, $L(\mathbf{e})$ takes one of two possible values, e.g. in $\{0, 1\}$. For example, a client of a financial institution requests a loan, but the classifier, on

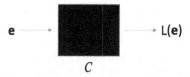

Fig. 2. A black-box classifier

the basis of his/her feature values (e.g. for EdLevel, Income, Age, etc.) assigns the label 1, for rejection. An explanation may be requested by the client, independently from the kind of classifier that is being used. The latter could be an explicit classification model, e.g. a classification tree or a logistic regression model. In these cases, we might be in a better position to given an explanation, because we can inspect the internals of the model [42]. However, we will put ourselves in the "worst scenario" in which we do not have access to the internal model. That is, we are confronted to a black-box classifier.

An approach to explanations that has become popular, specially in the absence of the model, assigns numerical *scores* to the feature values for an entity, trying to answer the question about which of the feature values contribute the most to the received label.

Example 20 Reusing a popular example from [38], let us consider the set of features $\mathcal{F} = \{$Outlook, Humidity, Wind$\}$, with $Dom($Outlook$) = \{$sunny, overcast, rain$\}$, $Dom($Humidity$) = \{$high, normal$\}$, $Dom($Wind$) = \{$strong, weak$\}$. An entity under classification has a value for each of the features, e.g. e $= ent($sunny, normal, weak$)$, and represents a particular weather condition. The problem consists in deciding about playing tennis or not under the conditions represented by that entity, which can be captured as a classification problem, with labels "yes" or "no".

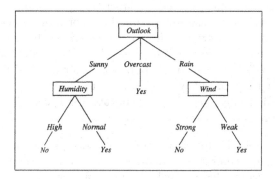

Fig. 3. A decision tree

In this case, the binary classifier is given as a decision-tree, as shown in Fig. 3. It could be displayed by double-clicking on the black box in Fig. 2. The decision is computed by following the feature values along the branches of the tree. The entity **e** at hand gets label yes.   □

Score-based methodologies are sometimes based on *counterfactual interventions*: What would happen with the label if we change this particular value, leaving the others fixed? Or the other way around: What if we leave this value fixed, and change the others? The resulting labels from these counterfactual interventions can be aggregated in different ways, leading to a score for the feature value under inspection.

A be more concrete, we can use the previous example, to detect and quantify the relevance (technically, the responsibility) of a feature value in **e** = *ent*(sunny, normal, weak), say for feature Humidity (underlined), by *hypothetically intervening* its value. In this case, if we change it from normal to high, we obtain a new entity **e′** = *ent*(sunny, high, weak), a *counterfactual version* of **e**. If we input this entity into the classifier, we now obtain the label no. This is an indication that the original feature value for Humidity is indeed relevant for the original classification.

In the next two sections we briefly introduce two scores. Both can be applied with open-box or black-box models. In both cases, we consider a finite set of features $\mathcal{F}$, with each feature $F \in \mathcal{F}$ having a finite domain, $Dom(F)$, where F, as function, takes its values. The features are applied to entities **e** in a population $\mathcal{E}$ of them. Actually, we identify the entity **e** with the record (or tuple) formed by the values the features take on it: $\mathbf{e} = \langle F_1(\mathbf{e}), \ldots, F_n(\mathbf{e}) \rangle$. Now, entities in $\mathcal{E}$ go through a binary *classifier*, C, that returns *labels* for them. We will assume the labels are 1 or 0. For example, the bank could have a classifier that automatically decides, for an entity, if it is worthy of a loan (0) or not (1).

10   The x-Resp Score

Assume that an entity **e** has received the label 1 by the classifier C, and we want to explain this outcome by assigning numerical scores to **e**'s feature values, in such a way, that a higher score for a feature value reflects that it has been important for the outcome. We do this now using the x-Resp score, whose definition we illustrate by means of an example (c.f. [10,11] for detailed treatments). For simplicity and for the moment, we will assume the features are also binary, i.e. they propositional, taking the values true or false (or 1 and 0, resp.) In Sect. 12, we consider a more general case.

Example 21 In Fig. 4, the black box is the classifier C. An entity **e** has gone through it obtaining label 1, shown in the first row in the figure. We want to assign a score to the feature value **x** for a feature $F \in \mathcal{F}$. We proceed, counterfactually, changing the value **x** into **x′**, obtaining a counterfactual version $\mathbf{e}_1$ of **e**. We classify $\mathbf{e}_1$, and we still get the outcome 1 (second row in the figure). In between, we may counterfactually change other feature values, **y, z** in **e**, into **y′, z′**, but keeping **x**, obtaining entity $\mathbf{e}_2$, and the outcome does not change (third

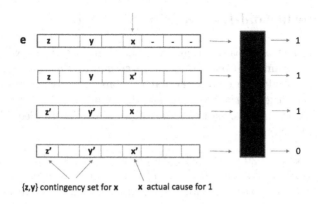

Fig. 4. Classified entity and its counterfactual versions

row). However, if we change in e_2, $\mathbf{x}$ into $\mathbf{x}'$, the outcome does change (fourth row).

This shows that the value $\mathbf{x}$ is relevant for the original output, but, for this outcome, it needs company, say of the feature values $\mathbf{y}, \mathbf{z}$ in $\mathbf{e}$. Proceeding as in actual causality as applied to tuples in a database in relation to query answering (c.f. Sect. 3.1), we can say that the feature value $\mathbf{x}$ in $\mathbf{e}$ is an *actual cause* for the classification, that needs a *contingency set* formed by the values $\mathbf{y}, \mathbf{z}$ in $\mathbf{e}$. In this case, the contingency set has size 2. If we found a contingency set for $\mathbf{x}$ of size 1 in $\mathbf{e}$, we would consider $\mathbf{x}$ even more relevant for the output. $\square$

On this basis, we can define [10,11]: (a) $\mathbf{x}$ is a *counterfactual explanation* for $L(\mathbf{e}) = 1$ if $L(\mathbf{e}\frac{\mathbf{x}}{\mathbf{x}'}) = 0$, for some $\mathbf{x}' \in Dom(F)$ (the domain of feature F). (Here we use the common notation $\mathbf{e}\frac{\mathbf{x}}{\mathbf{x}'}$ for the entity obtained by replacing $\mathbf{x}$ by $\mathbf{x}'$ in $\mathbf{e}$). (b) $\mathbf{x}$ is an *actual explanation* for $L(\mathbf{e}) = 1$ if there is a set of values $\mathbf{Y}$ in $\mathbf{e}$, with $\mathbf{x} \notin \mathbf{Y}$, and new values $\mathbf{Y}' \cup \{\mathbf{x}'\}$, such that $L(\mathbf{e}\frac{\mathbf{Y}}{\mathbf{Y}'}) = 1$ and $L(\mathbf{e}\frac{\mathbf{x}\mathbf{Y}}{\mathbf{x}'\mathbf{Y}'}) = 0$.

Contingency sets may come in sizes from 0 to $n - 1$ for feature values in records of length n. Accordingly, we can define for the actual cause $\mathbf{x}$: If $\mathbf{Y}$ is a minimum-size contingency set for $\mathbf{x}$, x-Resp$(\mathbf{x}) := \frac{1}{1+|\mathbf{Y}|}$; and as 0 when $\mathbf{x}$ is not an actual cause.

We will reserve the notion of counterfactual explanation *for (or counterfactual version of) an input entity* $\mathbf{e}$ *for any entity* $\mathbf{e}'$ *obtained from* $\mathbf{e}$ *by modifying feature values in* $\mathbf{e}$ *and that leads to a different label, i.e.* $L(\mathbf{e}) \neq L(\mathbf{e}')$. *Notice that from such an* $\mathbf{e}'$ *we can read off actual causes for* $L(\mathbf{e})$ *as feature values, and contingency sets for those actual causes. It suffices to compare* $\mathbf{e}$ *with* $\mathbf{e}'$.

In Sect. 11 we give a detailed example that illustrates these notions, and also show the use of ASPs for the specification and computation of counterfactual versions of a given entity, and the latter's x-Resp score.

11   Counterfactual-Intervention Programs

Together with illustrating the notions introduced in Sect. 10, we will introduce, by means of an example, *Counterfactual Intervention Programs* (CIPs). They are ASPs that specify the counterfactual versions of a given entity, and also, if so desired, only the *maximum-responsibility* counterfactual explanations, i.e. counterfactual versions that lead to a maximum x-Resp score. See [11] for many more details and examples.

Example 22 (Example 20 continued). We present now the CIP for the classifier based on the decision-tree, in *DLV-Complex* notation. We use annotation constants o, for "original entity", do, for "do a counterfactual intervention" (a single change of feature value), tr, for "entity in transition", and s, for "stop, the label has changed". We explain the program as we present it, and also by inserting comments in the *DLV* code.

Notice that after the facts, that include the domains and the input entity, we find the rule-based specification of the decision tree. The ent predicate, for "entity", uses an entity identifier (eid) in its first argument.

```
% facts:
    dom1(sunny). dom1(overcast). dom1(rain). dom2(high). dom2(normal).
    dom3(strong). dom3(weak).
    ent(e,sunny,normal,weak,o).    % original entity at hand

% specification of the decision-tree classifier:
    cls(X,Y,Z,1) :- Y = normal, X = sunny, dom1(X), dom3(Z).
    cls(X,Y,Z,1) :- X = overcast, dom2(Y), dom3(Z).
    cls(X,Y,Z,1) :- Z = weak, X = rain, dom2(Y).
    cls(X,Y,Z,0) :- dom1(X), dom2(Y), dom3(Z), not cls(X,Y,Z,1).

% transition rules: the initial entity or one affected by a value change
    ent(E,X,Y,Z,tr) :- ent(E,X,Y,Z,o).
    ent(E,X,Y,Z,tr) :- ent(E,X,Y,Z,do).

% counterfactual rule: alternative single-value changes
    ent(E,Xp,Y,Z,do) v ent(E,X,Yp,Z,do) v ent(E,X,Y,Zp,do) :-
                    ent(E,X,Y,Z,tr), cls(X,Y,Z,1), dom1(Xp), dom2(Yp),
                    dom3(Zp), X != Xp, Y != Yp, Z != Zp,
                    chosen1(X,Y,Z,Xp), chosen2(X,Y,Z,Yp),
                    chosen3(X,Y,Z,Zp).
```

In this rule's body we find the "choice operator". It is a predicate (to de defined next in the program), say $chosen_1(x, y, z, x')$, that, for each combination of values (x, y, z) "chooses" a single value for x'. This new value can be used to replace a value in the first argument of the entity. Similarly for $chosen_2(x, y, z, y')$ and $chosen_3(x, y, z, z')$. They can be defined by means of the next rules in the program [25].

```
% definitions of "chosen" predicates:
    chosen1(X,Y,Z,U) :- ent(E,X,Y,Z,tr), cls(X,Y,Z,1), dom1(U), U != X,
                        not diffchoice1(X,Y,Z,U).
    diffchoice1(X,Y,Z, U) :- chosen1(X,Y,Z, Up), U != Up, dom1(U).
    chosen2(X,Y,Z,U) :- ent(E,X,Y,Z,tr), cls(X,Y,Z,1), dom2(U), U != Y,
                        not diffchoice2(X,Y,Z,U).
    diffchoice2(X,Y,Z, U) :- chosen2(X,Y,Z, Up), U != Up, dom2(U).
    chosen3(X,Y,Z,U) :- ent(E,X,Y,Z,tr), cls(X,Y,Z,1), dom3(U), U != Z,
                        not diffchoice3(X,Y,Z,U).
    diffchoice3(X,Y,Z, U) :- chosen3(X,Y,Z, Up), U != Up, dom3(U).

% Not going back to initial entity (program constraint):
    :- ent(E,X,Y,Z,do), ent(E,X,Y,Z,o).
```

The last rule is a (hard) *program constraint* that avoids going back to the initial entity by performing value changes. This constraint makes the ASP evaluation engine discard those models where this happen [29].

```
% stop when label has been changed:
    ent(E,X,Y,Z,s) :- ent(E,X,Y,Z,do), cls(X,Y,Z,0).

% collecting changed values for each feature:
    expl(E,outlook,X)    :- ent(E,X,Y,Z,o), ent(E,Xp,Yp,Zp,s), X != Xp.
    expl(E,humidity,Y)   :- ent(E,X,Y,Z,o), ent(E,Xp,Yp,Zp,s), Y != Yp.
    expl(E,wind,Z)       :- ent(E,X,Y,Z,o), ent(E,Xp,Yp,Zp,s), Z != Zp.


    entAux(E) :- ent(E,X,Y,Z,s).         % auxiliary predicate to
                                         % avoid unsafe negation
                                         % in the constraint below
    :- ent(E,X,Y,Z,o), not entAux(E).    % discard models where
                                         % label does not change

% computing the inverse of x-Resp:
    invResp(E,M) :- #count{I: expl(E,I,_)} = M, #int(M), E = e.
```

The last rule returns, for a given entity, the number of values that have been changed in order to reach a counterfactual version of that entity. The inverse of this value can be used to compute a x-Resp score (the $\frac{1}{1+|\mathbf{Y}|}$ in Sect. 12).

Two counterfactual versions of **e** are obtained, as represented by the two essentially different stable models of the program, and determined by the atoms with the annotation **s** (below, we keep in them only the most relevant atoms, omitting initial facts and choice-related atoms):

```
{ent(e,sunny,normal,weak,o), cls(sunny,normal,strong,1),
 cls(sunny,normal,weak,1), cls(overcast,high,strong,1),
 cls(overcast,high,weak,1), cls(rain,high,weak,1),
 cls(overcast,normal,weak,1), cls(rain,normal,weak,1),
 cls(overcast,normal,strong,1), cls(sunny,high,strong,0),
 cls(sunny,high,weak,0), cls(rain,high,strong,0),
```

```
cls(rain,normal,strong,0), ent(e,sunny,high,weak,do),
ent(e,sunny,high,weak,tr), ent(e,sunny,high,weak,s),
expl(e,humidity,normal),invResp(e,1)}
```

```
{ent(e,sunny,normal,weak,o), cls(sunny,normal,strong,1),...,
cls(rain,normal,strong,0), ent(e,rain,normal,strong,do),
ent(e,rain,normal,strong,tr), ent(e,rain,normal,strong,s),
expl(e,outlook,sunny), expl(e,wind,weak), invResp(e,2)}
```

The first model shows the classifiers as a set of atoms, and, in its second last line, that ent(e,sunny,high,weak,s) is a counterfactual version (with label 0) of the original entity e, and is obtained from the latter by means of changes of values in feature Humidity, leading to an inverse score of 1. The second model shows a different counterfactual version of e, namely ent(e,rain,normal,strong,s), now obtained by changing values for features Outlook and Wind, leading to an inverse score of 2.

Let us now add, at the end of the program the following weak constraints:

```
% Weak constraints to minimize number of changes:         (*)
   :~ ent(E,X,Y,Z,o), ent(E,Xp,Yp,Zp,s), X != Xp.
   :~ ent(E,X,Y,Z,o), ent(E,Xp,Yp,Zp,s), Y != Yp.
   :~ ent(E,X,Y,Z,o), ent(E,Xp,Yp,Zp,s), Z != Zp.
```

If we run the program with them, the number of changes is minimized, and we basically obtain only the first model above, corresponding to the counterfactual entity $e' = ent(\text{sunny}, \text{high}, \text{weak})$. This is a maximum-responsibility counterfactual explanation.   □

As can be seen at the light of this example, more complex rule-based classifiers could be defined inside a CIP. It is also possible to invoke the classifier as an external predicate [11].

11.1   Bringing-in Domain Knowledge

The CIP-based specifications we have considered so far allow all kinds of counterfactual interventions on feature values. However, this may be undesirable or unrealistic in certain applications. For, example, we may not end up producing, and even less, using for score computation, some entities representing people who have the combination of values yes and yes for the propositional features Married and YoungerThan5. Declarative approaches to specification and computation of counterfactual explanations have the nice feature that *domain knowledge and semantic constraints* can be easily integrated with the base specification. Procedural approaches may, most likely, require changing the underlying code. We use an example to illustrate the point. For more details and a discussion see [11].

Example 23 (Example 22 continued). It could be that in a particular geographic region, "raining with a strong wind at the same time" is never possible. When producing counterfactual interventions for the entity e, such a combination should not be produced or considered.

This can be done by imposing a hard program constraint

```
% hard constraint disallowing a particular combination
   :- ent(E,rain,X,strong,tr).
```

that we add to the program in Example 22, from which we previously remove the weak constraints we had in (*) (in order not to discard any model for cardinality reasons). If we run the new program with *DLV*, we obtain only the first model in Example 22, corresponding to the counterfactual entity $\mathbf{e}' = ent(\mathsf{sunny}, \mathsf{high}, \mathsf{weak})$. □

12   The Generalized **Resp** Score

If we want to assign a numerical score to a feature value, say $v = F(\mathbf{e})$, where F has a relatively large domain, $Dom(F)$, it could be the case that counterfac-tually changing v into $v' \in Dom(F)$ changes the label (while leaving the other feature values fixed). However, it could be that for nearly all the other values in $Dom(F) \smallsetminus \{v, v'\}$, the label does not change. In this case, we might consider that maybe v is not such a strong reason for the originally obtained label, despite the fact that v is still a counterfactual explanation (with empty contingency set) according to Sect. 10.

For this reason, it might be better to consider all the possible alternative values for F, and define and compute the score in terms of an average of the label values, or an expected value for the label in case we have an underlying probability distribution P on the entity population $\mathcal{E}$. Such a general version of the x-**Resp** score was introduced and investigated in [9]. We briefly describe it starting with the simpler case of counterfactual explanations, i.e. without considering contingency sets. Next, we further generalize the score to consider the latter. So, in the following, the features do not have to be binary.

Assume that entity $\mathbf{e}$ has gone through a classifier and we have obtained label 1, which we would like to explain. Then, for a feature $F^\star \in \mathcal{F}$, we may consider as a score:

$$\mathsf{Counter}(\mathbf{e}, F^\star) := L(\mathbf{e}) - \mathbb{E}(L(\mathbf{e}') \mid \mathbf{e}'_{\mathcal{F} \smallsetminus \{F^\star\}} = \mathbf{e}_{\mathcal{F} \smallsetminus \{F^\star\}}). \tag{16}$$

Here, $\mathbf{e}_S$, for $S \subseteq \mathcal{F}$ is the entity $\mathbf{e}$ restricted to the features in S. This score measures the expected difference between the label for $\mathbf{e}$ and those for entities that coincide in feature values everywhere with $\mathbf{e}$ but on feature $F^\star$. Notice the essential counterfactual nature of this score, which is reflected in all the possible hypothetical changes of values for $F^\star$ in $\mathbf{e}$.

A problem with $\mathsf{Counter}$ is that changing a single value, no matter how, may not switch the original label, in which case no explanations are obtained. In order to address this problem, we can bring in *contingency sets* of feature values, which leads to the **Resp** score introduced in [9].

Again, consider $\mathbf{e} \in \mathcal{E}$, an entity under classification, for which $L(\mathbf{e}) = 1$, and a feature $F^\star \in \mathcal{F}$. Assume we have:

1. $\Gamma \subseteq \mathcal{F} \setminus \{F^\star\}$, a set of features that may end up accompanying feature $F^\star$.
2. $\bar{w} = (w_F)_{F \in \Gamma}$, $w_F \in Dom(F)$, $w_F \neq \mathbf{e}_F$, i.e. new values for features in Γ.
3. $\mathbf{e}' := \mathbf{e}[\Gamma := \bar{w}]$, i.e. reset $\mathbf{e}$'s values for Γ as in $\bar{w}$.
4. $L(\mathbf{e}') = L(\mathbf{e}) = 1$, i.e. there is no label change with $\bar{w}$ (but maybe with an extra change for $F^\star$, in next item).
5. There is $v \in Dom(F^\star)$, with $v \neq F^\star(\mathbf{e})$ and $\mathbf{e}'' := \mathbf{e}[\Gamma := \bar{w}, F^\star := v]$.

As in Sect. 10, if $L(\mathbf{e}) \neq L(\mathbf{e}'') = 0$, $F^\star(\mathbf{e})$ is an *actual causal explanation* for $L(\mathbf{e}) = 1$, with "contingency set" $\langle \Gamma, \mathbf{e}_\Gamma \rangle$, where $\mathbf{e}_\Gamma$ is the projection of $\mathbf{e}$ on Γ.

In order to define the "local" responsibility score, make v vary randomly under conditions 1.–5.:

$$\mathsf{Resp}(\mathbf{e}, F^\star, \Gamma, \bar{w}) := \frac{L(\mathbf{e}') - \mathbb{E}[L(\mathbf{e}'') \mid \mathbf{e}''_{\mathcal{F} \setminus \{F^\star\}} = \mathbf{e}'_{\mathcal{F} \setminus \{F^\star\}}]}{1 + |\Gamma|}. \qquad (17)$$

If, as so far, label 1 is what has to be explained, then $L(\mathbf{e}') = 1$, and the numerator is a number between 0 and 1. Here, Γ is fixed. Now, we can minimize its size, obtaining the (generalized) responsibility score as the maximum local value; everything relative to distribution P:

$$\mathsf{Resp}_{\mathbf{e}, F^\star}(F^\star(\mathbf{e})) := \quad max \quad \mathsf{Resp}(\mathbf{e}, F^\star, \Gamma, \bar{w}) \qquad (18)$$
$$|\Gamma| \; min., \; (18) > 0$$
$$\langle \Gamma, \bar{w} \rangle \models 1.-4.$$

This score was introduced in [9], where experiments and comparisons with other scores, namely Shap (c.f. Sect. 13) and the *FICO* score [18], are shown. Furthermore, different probability distributions are considered. Notice that, in order to compute this score, there is no need to access the internals of the classification model.

13   The **Shap** Score

In the context of classification, the Shapley value (c.f. Section 8) has taken the form of the Shap score [34], which we briefly introduce. Given the binary classifier, $\mathcal{C}$, on binary entities, it becomes crucial to identify a suitable game function. In this case, it will be expressed in terms of expected values (not unlike the causal-effect score), which requires an underlying probability space on the population of entities, $\mathcal{E}$. We will consider, to fix ideas, the *uniform probability space* on $\mathcal{E}$. Since we will consider only binary feature values, taking values 0 or 1, this is the uniform distribution on $\mathcal{E} = \{0, 1\}^n$, assigning probability $P^u(\mathbf{e}) = \frac{1}{2^n}$ to $\mathbf{e} \in \mathcal{E}$. One could consider other distributions [3, 9].

Given a set of features $\mathcal{F} = \{F_1, \ldots, F_n\}$, and an entity $\mathbf{e}$ whose label is to be explained, the set of players D in the game is $\mathcal{F}(\mathbf{e}) := \{F(\mathbf{e}) \mid F \in \mathcal{F}\}$, i.e. the set of feature values of $\mathbf{e}$. Equivalently, if $\mathbf{e} = \langle x_1, \ldots, x_n \rangle$, then $x_i = F_i(\mathbf{e})$. We assume these values have implicit feature identifiers, so that duplicates do not collapse, i.e. $|\mathcal{F}(\mathbf{e})| = n$. The game function is defined as follows. For $S \subseteq \mathcal{F}(\mathbf{e})$,

$$\mathcal{G}_{\mathbf{e}}(S) := \mathbb{E}(L(\mathbf{e}') \mid \mathbf{e}'_S = \mathbf{e}_S),$$

where $\mathbf{e}_S$: is the projection of $\mathbf{e}$ on S. This is the expected value of the label for entities $\mathbf{e}'$ when their feature values are fixed and equal to those in S for $\mathbf{e}$. Other than that, the feature values of $\mathbf{e}'$ may independently vary over $\{0,1\}$.

Now, one can instantiate the general expression for the Shapley value in (15), using this particular game function, as $Shapley(\mathcal{F}(\mathbf{e}), \mathcal{G}_{\mathbf{e}}, F(\mathbf{e}))$, obtaining, for a particular feature value $F(\mathbf{e})$:

$$\mathsf{Shap}(\mathcal{F}(\mathbf{e}), \mathcal{G}_{\mathbf{e}}, F(\mathbf{e})) := \sum_{S \subseteq \mathcal{F}(\mathbf{e}) \setminus \{F(\mathbf{e})\}} \frac{|S|!(n - |S| - 1)!}{n!} \times$$
$$(\mathbb{E}(L(\mathbf{e}'|\mathbf{e}'_{S \cup \{F(\mathbf{e})\}} = \mathbf{e}_{S \cup \{F(\mathbf{e})\}}) - \mathbb{E}(L(\mathbf{e}')|\mathbf{e}'_S = \mathbf{e}_S)).$$

Here, the label L acts as a Bernoulli random variable that takes values through the classifier. We can see that the Shap score is a weighted average of differences of expected values of the labels [34]. We may notice that counterfactual versions of the initial entity are implicitly considered.

The Shap score can be applied with black-box classifiers. Under those circumstances its computation takes exponential time in that all permutations of subsets of features are involved. However, sometimes, when the classifier is explicitly available, the computation cost can be brought down, even to polynomial time. This is the case for several classes of Boolean circuits that can be used as classifiers, and in particular, for decision trees [3,34,48]. For other explicit Boolean circuit-based classifiers, the computation of Shap is still #P-hard [3,48].

14   Final Remarks

Explainable data management and explainable AI (XAI) are effervescent areas of research. The relevance of explanations can only grow, as observed from- and due to the legislation and regulations that are being produced and enforced in relation to explainability, transparency and fairness of data management and AI/ML systems.

There are different approaches and methodologies in relation to explanations, with causality, counterfactuals and scores being prominent approaches that have a relevant role to play. Much research is still needed on the use of *contextual, semantic and domain knowledge*. Some approaches may be more appropriate in this direction, and we argue that declarative, logic-based specifications can be successfully exploited [11].

Still fundamental research is needed in relation to the notions of *explanation* and *interpretation*. An always present question is: *What is a good explanation?*. This is not a new question, and in AI (and other areas and disciplines) it has been investigated. In particular in AI, areas such as *diagnosis* and *causality* have much to contribute.

Now, in relation to *explanations scores*, there is still a question to be answered: *What are the desired properties of an explanation score?*. The question makes a lot of sense, and may not be beyond an answer. After all, the general Shapley value emerged from a list of *desiderata* in relation to coalition games, as the only measure that satisfies certain explicit properties [43,45]. Although the Shapley value is being used in XAI, in particular in its Shap incarnation, there could be a different and specific set of desired properties of explanation scores that could lead to a still undiscovered explanation score.

Acknowledgments. L. Bertossi has been a member of the Academic Network of RelationalAI Inc., where his interest in explanations in ML started. Part of this work was funded by ANID - Millennium Science Initiative Program - Code ICN17002. Help from Jessica Zangari and Mario Alviano with information about *DLV2*, and from Gabriela Reyes with the *DLV* program runs is much appreciated. We are grateful to an anonymous reviewer for valuable comments.

References

1. Arora, S., Barak, B.: Computational Complexity. Cambridge University Press, Cambridge (2009)
2. Arenas, M., Bertossi, L., Chomicki, J.: Consistent query answers in inconsistent databases. In: Proceedings of ACM PODS 1999, pp. 68–79 (1999)
3. Arenas, M., Pablo Barceló, P., Bertossi, L., Monet, M.: The tractability of SHAP-scores over deterministic and decomposable boolean circuits. In: Proceedings of AAAI 2021, pp. 6670–6678 (2021)
4. Bertossi, L.: Database repairing and consistent query answering. Synthesis Lect. Data Manag. **3**, 1–121 (2011)
5. Bertossi, L., Salimi, B.: From causes for database queries to repairs and model-based diagnosis and back. Theory Comput. Syst. **61**(1), 191–232 (2016). https://doi.org/10.1007/s00224-016-9718-9
6. Bertossi, L., Salimi, B.: Causes for query answers from databases: datalog abduction, view-updates, and integrity constraints. Int. J. Approx. Reason. **90**, 226–252 (2017)
7. Bertossi, L.: Repair-based degrees of database inconsistency. In: Balduccini, M., Lierler, Y., Woltran, S. (eds.) LPNMR 2019. LNCS, vol. 11481, pp. 195–209. Springer, Cham (2019). https://doi.org/10.1007/978-3-030-20528-7_15
8. Bertossi, L.: Specifying and computing causes for query answers in databases via database repairs and repair programs. Knowl. Inf. Syst. **63**(1), 199–231 (2021)
9. Bertossi, L., Li, J., Schleich, M., Suciu, D., Vagena, Z.: Causality-based explanation of classification outcomes. In: Proceedings of the Fourth Workshop on Data Management for End-To-End Machine Learning, DEEM@SIGMOD 2020, pp. 6:1–6:10 (2020)
10. Bertossi, L.: An ASP-based approach to counterfactual explanations for classification. In: Gutiérrez-Basulto, V., Kliegr, T., Soylu, A., Giese, M., Roman, D. (eds.) RuleML+RR 2020. LNCS, vol. 12173, pp. 70–81. Springer, Cham (2020). https://doi.org/10.1007/978-3-030-57977-7_5
11. Bertossi, L.: Declarative approaches to counterfactual explanations for classification. Publ. Theory Pract. Log. Program. (2021). https://doi.org/10.1017/S1471068421000582

12. Breiman, L., Friedman, J., Stone, C.J., Olshen, R.A.: Classification and Regression Trees. CRC Press, Boca Raton (1984)
13. Brewka, G., Eiter, T., Truszczynski, M.: Answer set programming at a glance. Commun. ACM **54**(12), 92–103 (2011)
14. Buneman, P., Khanna, S., Wang-Chiew, T.: Why and where: a characterization of data provenance. In: Van den Bussche, J., Vianu, V. (eds.) ICDT 2001. LNCS, vol. 1973, pp. 316–330. Springer, Heidelberg (2001). https://doi.org/10.1007/3-540-44503-X_20
15. Calimeri, F., Cozza, S., Ianni, G., Leone, N.: Computable functions in ASP: theory and implementation. In: Garcia de la Banda, M., Pontelli, E. (eds.) ICLP 2008. LNCS, vol. 5366, pp. 407–424. Springer, Heidelberg (2008). https://doi.org/10.1007/978-3-540-89982-2_37
16. Calimeri, F., Cozza, S., Ianni, G., Leone, N.: An ASP system with functions, lists, and sets. In: Erdem, E., Lin, F., Schaub, T. (eds.) LPNMR 2009. LNCS (LNAI), vol. 5753, pp. 483–489. Springer, Heidelberg (2009). https://doi.org/10.1007/978-3-642-04238-6_46
17. Caniupan, M., Bertossi, L.: The consistency extractor system: answer set programs for consistent query answering in databases. Data Knowl. Eng. **69**(6), 545–572 (2010)
18. Chen, C., Lin, K., Rudin, C., Shaposhnik, Y., Wang, S., Wang, T.: An interpretable model with globally consistent explanations for credit risk. CoRR, abs/1811.12615 (2018)
19. Chockler, H., Halpern, J.: Responsibility and Blame: a structural-model approach. J. Artif. Intell. Res. **22**, 93–115 (2004)
20. Dantsin, E., Eiter, T., Gottlob, G., Voronkov, A.: Complexity and expressive power of logic programming. ACM Comput. Surv. **33**(3), 374–425 (2001)
21. Deng, X., Papadimitriou, C.: On the complexity of cooperative solution concepts. Math. Oper. Res. **19**(2), 257–266 (1994)
22. Faigle, U., Kern, W.: The Shapley value for cooperative games under precedence constraints. Internat. J. Game Theory **21**, 249–266 (1992)
23. Gelfond, M., Lifschitz, V.: Classical negation in logic programs and disjunctive databases. N. Gener. Comput. **9**, 365–385 (1991)
24. Gelfond, M., Kahl, Y.: Knowledge Representation and Reasoning, and the Design of Intelligent Agents. Cambridge University Press, Press (2014)
25. Giannotti, F., Greco, S., Sacca, D., Zaniolo, C.: Programming with non-determinism in deductive databases. Ann. Math. Artif. Intell. **19**(1–2), 97–125 (1997)
26. Halpern, J., Pearl, J.: Causes and explanations: a structural-model approach. Part I: causes. Br. J. Philos. Sci. **56**(4), 843–887 (2005)
27. Halpern, J.Y.: A modification of the Halpern-Pearl definition of causality. In: Proceedings of IJCAI 2015, pp. 3022–3033 (2015)
28. Hunter, A., Konieczny, S.: On the measure of conflicts: Shapley inconsistency values. Artif. Intell. **174**(14), 1007–1026 (2010)
29. Leone, N., et al.: The DLV system for knowledge representation and reasoning. ACM Trans. Comput. Log. **7**(3), 499–562 (2006)
30. Livshits, E., Bertossi, L., Kimelfeld, B., Sebag, M.: The Shapley value of tuples in query answering. In: Proceedings of ICDT 2020, pp. 20:1–20:19 (2020)
31. Livshits, E., Kimelfeld, B.: The Shapley value of inconsistency measures for functional dependencies. In: Proceedings of ICDT 2021, pp. 15:1–15:19 (2021)

32. Lopatenko, A., Bertossi, L.: Complexity of consistent query answering in databases under cardinality-based and incremental repair semantics. In: Schwentick, T., Suciu, D. (eds.) ICDT 2007. LNCS, vol. 4353, pp. 179–193. Springer, Heidelberg (2006). https://doi.org/10.1007/11965893_13

33. Lucic, A., Haned, H., de Rijke, M.: Explaining predictions from tree-based boosting ensembles. CoRR, abs/1907.02582 (2019)

34. Lundberg, S., et al.: From local explanations to global understanding with explainable AI for trees. Nat. Mach. Intell. **2**(1), 2522–5839 (2020)

35. Lundberg, S., Lee, S.: A unified approach to interpreting model predictions. In: Proceedings of Advances in Neural Information Processing Systems, pp. 4765–4774 (2017)

36. Meliou, A., Gatterbauer, W., Moore, K.F., Suciu, D.: The complexity of causality and responsibility for query answers and non-answers. In: Proceedings of VLDB 2010, pp. 34–41 (2010)

37. Meliou, A., Gatterbauer, W., Halpern, J.Y., Koch, C., Moore, K.F., Suciu, D.: Causality in databases. IEEE Data Eng. Bull. **33**(3), 59–67 (2010)

38. Mitchell, T.M.: Machine Learning. McGraw-Hill, New York (1997)

39. Molnar, C.: Interpretable machine learning: a guide for making black box models explainable (2020). https://christophm.github.io/interpretable-ml-book

40. Nisan, N., Roughgarden, T., Tardos, E., Vazirani, V.V. (eds.): Algorithmic Game Theory. Cambridge University Press, Cambridge (2007)

41. Reshef, A., Kimelfeld, B., Livshits, E.: The impact of negation on the complexity of the Shapley value in conjunctive queries. In: Proceedings of PODS 2020, pp. 285–297 (2020)

42. Rudin, C.: Stop explaining black box machine learning models for high stakes decisions and use interpretable models instead. Nat. Mach. Intell. **1**, 206–215 (2019). arXiv:1811.10154 (2018)

43. Roth, A.E. (ed.): The Shapley Value: Essays in Honor of Lloyd S. Shapley. Cambridge University Press, Cambridge (1988)

44. Salimi, B., Bertossi, L., Suciu, D., Van den Broeck, G.: Quantifying causal effects on query answering in databases. In: Proceedings of 8th USENIX Workshop on the Theory and Practice of Provenance (TaPP) (2016)

45. Shapley, L.S.: A value for n-person games. Contrib. Theory Games **2**(28), 307–317 (1953)

46. Struss, P.: Model-based problem solving. In: Handbook of Knowledge Representation, chap. 4, pp. 395–465. Elsevier (2008)

47. Suciu, D., Olteanu, D., Re, C., Koch, C.: Probabilistic Databases. Synthesis Lectures on Data Management. Morgan & Claypool (2011)

48. Van den Broeck, G., Lykov, A., Schleich, M., Suciu, D.: On the tractability of SHAP explanations. In: Proceedings of AAAI 2021, pp. 6505–6513 (2021)

Author Index

Printed in the United States
by Baker & Taylor Publisher Services

Printed in the United States
by Baker & Taylor Publisher Services